Exploring Deepfakes

Deploy powerful AI techniques for face replacement and more with this comprehensive guide

Bryan Lyon

Matt Tora

BIRMINGHAM—MUMBAI

Exploring Deepfakes

Group Product Manager: Ali Abidi

Publishing Product Manager: Gebin George, Sunith Shetty

Senior Editor: David Sugarman

Technical Editor: Kavyashree K. S.

Copy Editor: Safis Editing

Project Coordinator: Farheen Fathima

Proofreader: Safis Editing

Indexer: Sejal Dsilva

Production Designer: Joshua Misquitta

Marketing Coordinator: Shifa Ansari

First published: February 2023

Production reference: 1280223

Published by Packt Publishing Ltd.

Livery Place

35 Livery Street

Birmingham

B3 2PB, UK.

ISBN 978-1-80181-069-2

www.packtpub.com

Contributors

About the authors

Bryan Lyon is a seasoned AI expert with over a decade of experience in and around the field. His background is in computational linguistics and he has worked with the cutting-edge open source deepfake software Faceswap since 2018. Currently, Bryan serves as the chief technology officer for an AI company based in California.

Matt Tora is a seasoned software developer with over 15 years of experience in the field. He specializes in machine learning, computer vision, and streamlining workflows. He leads the open source deepfake project Faceswap and consults with VFX studios and tech start-ups on integrating machine learning into their pipelines.

About the reviewer

Saisha Chhabria is a computer engineer with diverse experience ranging from software engineering to deep learning. She is currently based in Singapore and is pursuing her master's in computing, specializing in artificial intelligence from the National University of Singapore. She strives for challenging opportunities to build upon her repertoire of computation, development, and collaboration skills, and aspires to combat problems that impact the community.

Table of Contents

3

Acquiring and Processing Data   25

4

The Deepfake Workflow   35

Part 2: Getting Hands-On with the Deepfake Process

Part 3: Where to Now?

8

Applying the Lessons of Deepfakes   139

9

The Future of Generative AI   151

Preface

The media attention around deepfakes often focuses on the ills and dangers of the technology. Even many of the articles about what deepfakes can do fail to account for why you might want to do it. The truth is that deepfakes bring new techniques and abilities to neural networks for those who know how to use them.

Beyond replacing faces, deepfakes provide insights into all areas of generative AI, especially when traditional methods fall short. Join us as we explore what deepfakes are, what they can be used for, and how they may change in the future.

A manifesto for the ethical use of deepfakes

There are a lot of concerns when it comes to the use of deepfakes. To that end, we have to establish some common ground so that we can communicate and discuss deepfake technology:

- Deepfakes are not for creating inappropriate content
- Deepfakes are not for changing faces without consent or with the intent of hiding their use
- Deepfakes are not to be utilized for any illicit, unethical, or questionable purposes
- Deepfakes exist to experiment with and discover AI techniques, for social or political commentary, movies, and any number of other ethical and reasonable uses

We are very troubled by the fact that Faceswap has been used in unethical and disreputable ways. However, we support the development of tools and techniques that can be used ethically, as well as providing education and experience in AI for anyone who wants to learn it hands-on. We take a zero-tolerance approach to anyone using Faceswap for any unethical purposes and actively discourage any such uses.

Who this book is for

This book is for anyone interested in learning about deepfakes. From academics to researchers to content creators to developers, we've written this book so it has something for everybody.

The early chapters will cover the essential background of deepfakes, how they work, their ethics, and how to make one yourself using free software that you can download and use without any technical knowledge.

The middle chapters will go in depth into the exact methodology that deepfakes use to work, including working code that you can run and follow step-by-step as we get hands-on with the major processes of deepfakes: extraction, training, and conversion.

The final chapters will look at where you can go from there. They cover how to use deepfake techniques in your own tasks, and where the technology might go in the future.

What this book covers

Chapter 1, *Surveying Deepfakes*, provides a look into the past and present of deepfakes with a description of how they work and are used.

Chapter 2, *Examining Deepfake Ethics and Dangers*, provides a look at the sordid history of deepfakes and guidelines on creating ethical deepfakes.

Chapter 3, *Mastering Data*, teaches you how to get the most from your data, whether you make it yourself or have to find it.

Chapter 4, *The Deepfake Workflow*, provides a step-by-step walk-through of using Faceswap from the installation to the final output.

Chapter 5, *Extracting Faces*, is where we begin our hands-on dive into the code of a deepfake by learning how we detect, align, and mask faces for deepfakes.

Chapter 6, *Training a Deepfake Model*, is where we continue exploring the code as we train a model from scratch, including defining the layers of the model, feeding it images, and updating the model weights.

Chapter 7, *Swapping the Face Back into the Video*, is where we complete the code analysis with conversion, the process that puts the swapped face back into the original video.

Chapter 8, *Applying the Lessons of Deepfakes*, teaches you the process of solving hypothetical problems using deepfake techniques.

Chapter 9, *The Future of Generative AI*, examines where generative AI will move in the future and what limitations they need to overcome.

To get the most out of this book

This book is designed to build knowledge as you read through the chapters. If you're starting with no background knowledge of deepfakes, then we suggest you start at the beginning. If you want to skip straight to the code, then you'll want to look at *Part 2* (though we hope you'll give *Part 1* a peruse once you're ready). If you only care about what you can do with the techniques moving forward, then check out *Part 3* (but I promise that the earlier parts have some juicy nuggets of information).

We use Python for all code examples in this book. If you know Python, you should be able to understand all the code samples with the help of the text. If you don't know Python, then don't worry! There is a lot of non-code explanation, and even the code includes hands-on explanations of what is going on in it.

All the libraries used in this book are explained when they're used, but this book should not be considered a guide or in-depth explanation of any of the libraries. Many of these libraries have books of their own dedicated to them, and their use in this book is solely functional.

Software covered in the book		Operating system requirements
Python	Faceswap	Windows, macOS, or Linux
PyTorch	OpenCV	
Pillow (PIL Fork)		

We use Anaconda (https://www.anaconda.com/) for package management and sandboxing throughout this book. If you want to follow along, we highly recommend you install it from the site listed here. If you would rather use Python virtual environments, you may, but if you do, the instructions in this book will not always work without modification, especially installing the necessary packages. If you choose to use that route, you will have to find the correct version of libraries to install yourself.

If you are using the digital version of this book, we advise you access the code from the book's GitHub repository (a link is available in the next section). Doing so will help you avoid any potential errors related to the copying and pasting of code.

Included in each hands-on chapter is a list of exercises. Please don't take these as directions on what you must do, but consider them as helpers to more completely understand what it is that the code is doing and how you can use the techniques for yourself. They do not have "answers" as they are not really questions; they're just prompts for you to find new and exciting ways to apply your knowledge.

If you do complete any of the exercises (or come up with something impressive of your own), we'd appreciate it if you would "fork" the book's repo into your own GitHub account and show the world your accomplishment! We'd love to see what you can do with deepfakes.

Download the example code files

You can download the example code files for this book from GitHub at https://github.com/PacktPublishing/Exploring-Deepfakes. If an update to the code, it will be updated in the GitHub repository. there's

We also have other code bundles from our rich catalog of books and videos available at https://github.com/PacktPublishing/. Check them out!

Conventions used

There are a number of text conventions used throughout this book.

`Code in text`: Indicates code words in text, database table names, folder names, filenames, file extensions, pathnames, dummy URLs, user input, and Twitter handles. Here is an example: "Mount the downloaded `WebStorm-10*.dmg` disk image file as another disk in your system."

A block of code is set as follows:

```
html, body, #map {
  height: 100%;
  margin: 0;
  padding: 0
}
```

When we wish to draw your attention to a particular part of a code block, the relevant lines or items are set in bold:

```
[default]
exten => s,1,Dial(Zap/1|30)
exten => s,2,Voicemail(u100)
exten => s,102,Voicemail(b100)
exten => i,1,Voicemail(s0)
```

Any command-line input or output is written as follows:

```
$ mkdir css
$ cd css
```

Bold: Indicates a new term, an important word, or words that you see onscreen. For instance, words in menus or dialog boxes appear in **bold**. Here is an example: "Select **System info** from the **Administration** panel."

> **Tips or important notes**
> Appear like this.

Get in touch

Feedback from our readers is always welcome.

General feedback: If you have questions about any aspect of this book, email us at `customercare@packtpub.com` and mention the book title in the subject of your message.

Errata: Although we have taken every care to ensure the accuracy of our content, mistakes do happen. If you have found a mistake in this book, we would be grateful if you would report this to us. Please visit `www.packtpub.com/support/errata` and fill in the form.

Piracy: If you come across any illegal copies of our works in any form on the internet, we would be grateful if you would provide us with the location address or website name. Please contact us at `copyright@packt.com` with a link to the material.

If you are interested in becoming an author: If there is a topic that you have expertise in and you are interested in either writing or contributing to a book, please visit `authors.packtpub.com`.

Share your thoughts

Once you've read *Exploring Deepfakes*, we'd love to hear your thoughts! Scan the QR code below to go straight to the Amazon review page for this book and share your feedback.

https://packt.link/r/1-801-81069-9

Your review is important to us and the tech community and will help us make sure we're delivering excellent quality content.

Download a free PDF copy of this book

Thanks for purchasing this book!

Do you like to read on the go but are unable to carry your print books everywhere?

Is your eBook purchase not compatible with the device of your choice?

Don't worry, now with every Packt book you get a DRM-free PDF version of that book at no cost.

Read anywhere, any place, on any device. Search, copy, and paste code from your favorite technical books directly into your application.

The perks don't stop there, you can get exclusive access to discounts, newsletters, and great free content in your inbox daily

Follow these simple steps to get the benefits:

1. Scan the QR code or visit the link below

https://packt.link/free-ebook/9781801810692

2. Submit your proof of purchase
3. That's it! We'll send your free PDF and other benefits to your email directly

Part 1:
Understanding Deepfakes

Deepfakes are a new (and controversial) technique using generative AI. But despite the basic idea that they swap one face with another, how much do you really know about deepfakes?

It's normal to have questions about something such as deepfakes, and this section will address those questions. We'll start with the basics of how they work and the machine learning principles that deepfakes are built on, and then take a look at what software is available to make deepfakes. After that, we'll examine the ethics of deepfakes, including the unsavory beginnings, and build a framework of sorts to evaluate how to make ethical deepfakes. Then, we'll look at the most important part of creating a deepfake: data, including explanations of what makes good data and how to get the most from your (not-so-great) data. Finally, we'll walk you through a complete deepfake using Faceswap, an open source deepfake program.

By the end of this part, you'll have a good understanding of what makes a deepfake, how they work, and even how to make one for yourself.

This part comprises the following chapters:

- *Chapter 1, Surveying Deepfakes*
- *Chapter 2, Examining Deepfake Ethics and Dangers*
- *Chapter 3, Acquiring and Processing Data*
- *Chapter 4, The Deepfake Workflow*

1
Surveying Deepfakes

Understanding deepfakes begins with understanding where they came from and what they can do. In this chapter, we'll begin to explore deepfakes and their operation. We will go through the basics of what makes a deepfake work, talking about the differences between a **generative auto-encoder** and a **generative adversarial network (GAN)**. We will examine their usTo PD: es in media, education, and advertising. We'll investigate their limitations and consider how to plan and design your deepfakes to avoid the common pitfalls. Finally, we'll examine existing deepfake software and discuss what each kind can do.

We'll cover this in the following sections:

- Introducing deepfakes
- Exploring the uses of deepfakes
- Discovering how deepfakes work
- Assessing the limitations of generative AI
- Looking at existing deepfake software

Introducing deepfakes

The name **deepfake** comes from a portmanteau of "deep", referring to **deep learning**, and "fake," referring to the fact that the images generated are not genuine. The term first came into use on the popular website Reddit, where the original author released several deepfakes of adult actresses with other women's faces artificially applied to them.

> **Note**
> The ethics of deepfakes are controversial, and we will cover this in more depth in *Chapter 2, Examining Deepfake Ethics and Dangers*.

This unethical beginning is still what the technology is most known for, but it's not all that it can be used for. Since that time, deepfakes have moved into movies, memes, and more. Tom Cruise signed up for Instagram only after "Deep Tom Cruise" beat him to it. Steve Buscemi has remarked to Stephen Colbert that he "never looked better" when his face was placed on top of Jennifer Lawrence's and a younger version of Bill Nighy was deepfaked onto his own older self for a news clip from the "past" in the movie *Detective Pikachu*.

In this book, we will be taking a fairly narrow view of what deepfaking is, so let's define it now. A deepfake is the use of a **neural network** trained on two faces to replace one face with another. There are other technologies to swap faces that aren't deepfakes, and there are generative AIs that do other things besides swapping faces but to include all of those in the term just muddies the water and confuses the issue.

Exploring the uses of deepfakes

The original use of Deepfakes might be the one that required the least amount of imagination. Putting one person's face on another's person has many different uses in various fields. Please don't consider the ideas here as the full extent of the capabilities of deepfakes – someone is bound to imagine something new!

Entertainment

Entertainment is the first area that comes to mind for most people when they consider the usage of deepfakes. There are two main areas of entertainment in which I see deepfakes playing a significant role: *narrative* and *parody*.

Narrative

The utility of deepfakes in movies is obvious. Imagine an actor's face being superimposed onto their stunt double or an actor who becomes unavailable being replaced by another performer without any changes to the faces in the final movie.

While deepfakes may not seem good enough, deepfakes are already being used in Hollywood and other media today – from *Detective Pikachu*, which used deepfakes to de-age Bill Nighy, to *For All Mankind*, which used it to put actors face to face with Ronald Reagan. Agencies and VFX shops are all examining how to use deepfakes in their work.

These techniques are not unique to deepfakes. CGI (in this book, referring to 3D graphics) face replacements have been used in many movies. However, using CGI face replacement is expensive and complicated, requiring filming to be done in particular ways with lots of extra data captured to be used by the artists to get the CGI face to look good in the final scene. This is an art more than a science and requires extensive skills and knowledge to accomplish. Deepfakes solve many of these problems making new forms of face replacements possible.

Making a deepfake requires no special filming techniques (although some awareness will make the process smoother). Deepfakes also require very little attention or skill compared to CGI face replacements. This makes it ideal for lower-cost face replacements, but it can also be higher-quality since the AI accounts for details that even the most dedicated artist can't recreate.

Parody

Parody is an extremely popular form of social criticism and forms the basis for entire To PD: movies, TV shows, and other forms of media. Parody is normally done by professional impersonators. In some cases, those impersonators look (or can be made to look) similar to the person they're impersonating. Other times, there is a reliance on their performance to make the impersonation clear.

Deepfakes provide an opportunity to change the art of parody wherein the impersonator can be made to look like the individual being parodied via a deepfake instead of by chance of birth. By removing the attention from basic appearance, deepfakes allow the focus to be placed directly on the performance itself.

Deepfakes also enable a whole new form of parody in which normal situations can become parodic simply due to the changed face. This particular form becomes humorous due to the distinct oddity of very different faces, instead of an expected swap.

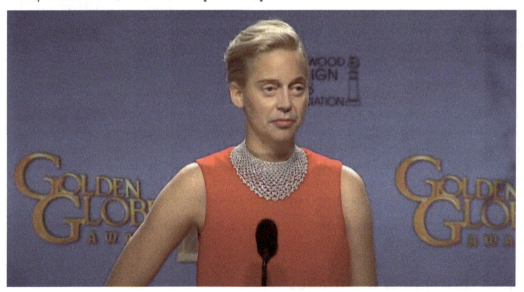

Figure 1.1 – Steve Buscemi as Jennifer Lawrence by birbfakes

> **Note**
> This image is included with the kind permission of its original creator, birbfakes. You can view the original video here: `https://youtu.be/r1jng79a5xc`.

Video games

Video games present an interesting opportunity when it comes to deepfakes. The idea here is that a computer-generated character could be deepfaked into a photorealistic avatar. This could be done for any character in the game, even the player's character. For example, it would be possible to make a game in which, when the player's character looked into a mirror, they would see their own face looking back at them. Another possibility would be to replace a non-player character with a deepfake of the original actor, allowing for a far more realistic appearance without making a complete 3D clone of the actor.

Education

Education could also benefit from deepfakes. Imagine if your history class had a video of Abraham Lincoln himself reading the Gettysburg address. Or a corporate training video in which the entire video is hosted by the public mascot (who may not even be a real person) without having to resort to costumes or CGI. It could even be used to allow multiple videos or scenes filmed at significantly different times to appear to be more cohesive by appearing to show the actor at the same time.

Many people are very visual learners and seeing a person "come alive" can really bring the experience home. Bringing the pre-video past to life using deepfakes enables a whole new learning experience. One example of this is the Dalí Museum, which created a series of videos of Salvador Dalí talking to guests. This was done by training a deepfake model on an actor to put Dalí's face on the videos. Once the model was trained and set up, they were able to convert many videos, saving a lot of time and effort compared to a CGI solution.

Advertisements

Advertising agencies are always looking for the newest way to grab attention and deepfakes could be a whole new way to catch viewers' attention. Imagine if you walked past a clothing store, you stopped to look at an item of clothing in the window, and suddenly the screen beside the item showed a video of an actor wearing the item but with your face, allowing you to see how the item would look on you. Alternatively, a mascot figure could be brought to life in a commercial. Deepfakes offer a whole new tool for creative use, which can grab attention and provide whole new experiences in advertising.

Now that we've got some idea of a few potential uses for deepfakes, let's take a quick look under the hood and see how they work.

Discovering how deepfakes work

Deepfakes are a unique variation of a generative auto-encoder being used to generate the face swap. This requires a special structure, which we will explain in this section.

Generative auto-encoders

The particular type of neural network that regular deepfakes use is called a **generative auto-encoder**. Unlike a **Generative Adversarial Network (GAN)**, an auto-encoder does not use a discriminator or any "adversarial" techniques.

All auto-encoders work by training a collection of neural network models to solve a problem. In the case of generative auto-encoders, the AI is used to generate a new image with new details that weren't in the original image. However, with a normal auto-encoder, the problem is usually something such as **classification** (deciding what an image is), **object identification** (finding something inside an image), or **segmentation** (identifying different parts of an image). To do this, there are two types of models used in the autoencoder – the **encoder** and **decoder**. Let's see how this works.

The deepfake training cycle

The training cycle is a cyclical process in which the model is continuously trained on images until stopped. The process can be broken down into four steps:

- **Encode** faces into smaller intermediate representations.
- **Decode** the intermediate representations back into faces.
- Calculate the **loss** of (meaning, the difference between) the original face and the output of the model.
- Modify (**backpropagate**) the models toward the correct answer.

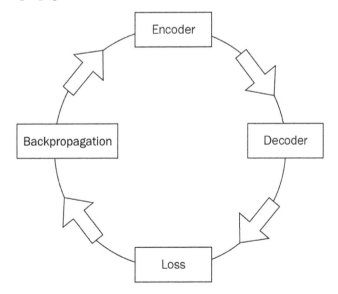

Figure 1.2 – Diagram of the training cycle

In more detail, the process unfolds as follows:

- The **encoder's** job is to **encode** two different faces into an array, which we call the intermediate representation. The intermediate representation is much smaller than the original image size, with enough space to describe the lighting, pose, and expression of the faces. This process is similar to compression, where unnecessary data is thrown out to fit the data into a smaller space.

- The **decoder** is actually a matched pair of models, which turn the intermediate representation back into faces. There is one decoder for each of the input faces, which is trained only on images of that one person's face. This process tries to create a new face that matches the original face that was given to the encoder and encoded into the intermediate representation.

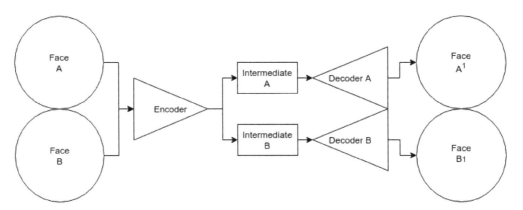

Figure 1.3 – Encoder and decoder

- **Loss** is a score that is given to the auto-encoder based on how well it recreates the original faces. This is calculated by comparing the original image to the output from the encoder-decoder process. This comparison can be done in many ways, from a strict difference between them or something significantly more complicated that includes human perception as part of the calculation. No matter how it's done, the result is the same: a number from 0 to 1, with 0 being the score for the model returning the exact same image and 1 being the exact opposite or the image. Most of the numbers will fall between 0 to 1. However, a perfect reconstruction (or its opposite) is impossible.

> **Note**
>
> The loss is where an auto-encoder differs from a GAN. In a GAN, the comparison loss is either replaced or supplemented with an additional network (usually an auto-encoder itself), which then produces a loss score of its own. The theory behind this structure is that the loss model (called a discriminator) can learn to get better at detecting the output of the generating model (called a generator) while the generator can learn to get better at fooling the discriminator.

- Finally, there is **backpropagation**, a process in which the models are adjusted by following the path back through both the decoder and encoder that generated the face and nudging those paths toward the correct answer.

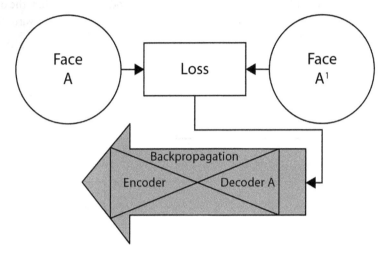

Figure 1.4 – Loss and backpropagation

Once complete, the whole process starts back over at the encoder again. This continues to repeat until the neural network has finished training. The decision of when to end training can happen in several ways. It can happen when a certain number of repetitions have occurred (called **iterations**), when all the data has been gone through (called an **epoch**), or when the results meet a certain loss score.

Why not GANs?

GANs are one of the current darlings of generative networks. They are extremely popular and used extensively, being used particularly for super-resolution (intelligent upscaling), music generation, and even sometimes deepfakes. However, there are some reasons that they're not used in all deepfake solutions.

GANs are popular due to their "imaginative" nature. They learn through the interaction of their generator and discriminator to fill in gaps in the data. Because they can fill in missing pieces, they are great at reconstruction tasks or at tasks where new data is required.

The ability of a GAN to create new data where it is missing is great for numerous tasks, but it has a critical flaw when used for deepfakes. In deepfakes, the goal is to replace one face with another face. An imaginative GAN would likely learn to fill the gaps in the data from one face with the data from the other. This leads to a problem that we call "identity bleed" where the two faces aren't swapped properly; instead, they're blended into a face that doesn't look like either person, but a mix of the two.

This flaw in a GAN-created deepfake can be corrected or prevented but requires much more careful data collection and processing. In general, it's easier to get a full swap instead of a blending by using a generative auto-encoder instead of a GAN.

The auto-encoder structure

Another name for an auto-encoder is an "hourglass" model. The reason for this is that each layer of an encoder is smaller than the layer before it while each layer of a decoder is larger than the one before. Because of this, the auto-encoder figure starts out large at the beginning, shrinks toward the middle, and then widens back out again as it reaches the end:

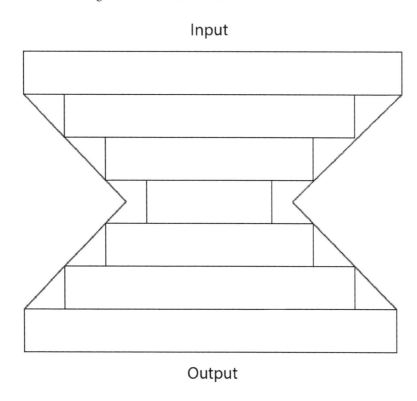

Figure 1.5 – Hourglass structure of an autoencoder

While these methods are flexible and have many potential uses, there are limitations. Let's examine those limitations now.

Assessing the limitations of generative AI

Generative AIs like those used in deepfakes are not a panacea and actually have some significant limitations. However, by knowing about these limitations, they can generally be worked around or sidestepped with careful design.

Resolution

Deepfakes are limited in the resolution that they can swap. This is a hardware and time limitation: greater hardware and more time can provide higher resolution swaps. However, this is not a 1:1 linear growth. Doubling the resolution (from, say, 64x64 to 128x128) actually quadruples the amount of required **VRAM** – that is, the memory that a GPU has direct access to – and the time necessary to train is expanded a roughly equivalent amount. Because of this, resolution is often a balancing act, where you'll want to make the deepfake the lowest resolution you can without sacrificing the results.

Training required for each face pair

To provide the best results, traditional deepfakes require that you train on every face pair that you wish to swap. This means that if you wanted to swap your own face with two of your friends, you'd have to train two separate models. This is because each model has one encoder and two decoders, which are trained only to swap the faces they were given.

There is a workaround to some multi-face swaps. In order to swap additional faces, you could write your own version with more than two decoders allowing you to swap additional faces. This is an imperfect solution, however, as each decoder takes up a significant amount of VRAM, requiring you to balance the number of faces carefully.

It may be better to simply train multiple pairs. By splitting the task up on multiple computers, you could train multiple models simultaneously, allowing you to create many face pairs at once.

Another option is to use a different type of AI face replacement. **First Order Model** (which is covered in the *Looking at existing deepfake software* section of this chapter) uses a different technique: instead of a paired approach, it uses AI to animate an image to match the actions of a replacement. This solution removes the need to retrain on each face pair, but comes at the cost of greatly reduced quality of the swap.

Training data

Generative AIs requires a significant amount of training data to accomplish their tasks. Sometimes, finding sufficient data or data of a high-enough quality is not possible. For example, how would someone create a deepfake of William Shakespeare when there are no videos or photographs of him? This is a tricky problem but can be worked around in several ways. While it is unfortunately impossible to create a proper deepfake of England's greatest playwright, it would be possible to use an actor who looks like his portraits and then deepfake that actor as Shakespeare.

> Tip
> We will cover more on how to deal with poor or insufficient data in *Chapter 3, Mastering Data*.

Finding sufficient data (or clever workarounds) is the most difficult challenge that any data scientist faces. Occasionally, there simply is no way to get sufficient data. This is when you might need to re-examine the video to see whether there is another way to shoot it to avoid the lack of data, or you might try using other sources of similar data to patch the gaps. Sometimes, just knowing the limitations in advance can prevent a problem – other times, a workaround in the last minutes may be enough to save a project from failure.

While everyone should know the data limitations, knowing the limitations of the process is only for experts. If you are only looking to use deepfakes, you'll probably use existing software. Let's explore those next.

Looking at existing deepfake software

There have been many programs that have risen to fill the niche of deepfaking; however, few of them are still under development or supported. The rapid development of GPU hardware and AI software has led to unique challenges in software development, and many deepfake programs are no longer usable. However, there are still several deepfake software programs and, in this section, we'll go over the major options.

> **Important Note**
> The authors have made every effort to be unbiased in this section, but are among the developers of Faceswap. Faceswap will be covered in more detail in *Chapter 4, The Deepfake Workflow*, with a walkthrough of the workflow of a deepfake through the Faceswap software.

Faceswap

Faceswap is a **Free and Open Source** (**FOSS**) software program for creating deepfakes. It's released under the GPL3 and can be used by anyone anywhere. It's written in Python and runs AI on the TensorFlow backend. It supports NVIDIA, AMD, and Apple GPUs for accelerating the machine learning models, or can be run on a CPU at a reduced speed. There are installers for Windows and Linux that can help by installing all the needed libraries and tools inside of a self-contained environment.

It's available at `https://Faceswap.dev/`.

DeepFaceLab

Originally a fork of Faceswap, DeepFaceLab is now developed mostly by Ivan Perov. DeepFaceLab is another FOSS software program for deepfakes. It is known for more experimental models and features. There is no GUI, but there are Jupyter Notebooks that can be run in any of the Jupyter environments. There is also a DirectML version, which provides another option for people using Windows. There are fully contained builds that are packaged together into a single compressed file, which provides a fully working package for many operating systems.

It's available at `https://github.com/iperov/DeepFaceLab`.

First Order Model

First Order Model works in a fundamentally different way from Faceswap and DeepFaceLab. Instead of swapping a face onto a new video, First Order Model "puppets" the face, making it match the movements of a video but leaving the face the same. Furthermore, it doesn't require training on each face pair, making it easy to use to make quick deepfakes where you can "animate" a person even with just a single photo of them.

It is important to note that while the First Order Model software is available freely, it is licensed only for non-commercial use: if you want to use it in a commercial context, you'll need to contact the author for a license. It's available at `https://github.com/AliaksandrSiarohin/first-order-model`.

Reface

Reface is yet another method of creating deepfakes. Reface is closed source and proprietary, so we can't analyze exactly how it works, but it uses a zero-shot learning method like First Order Model to swap faces without requiring training on a pair of swaps. Reface offers apps for Apple iOS and Android and does the swap in the cloud, making it easier to get a quick result, but means that you might not be able to swap the exact clip you want, and licensing may be an issue.

It's available at `https://reface.ai/`.

Summary

The technology of deepfakes is not itself anything new or unique. These techniques existed in various forms long before they were applied to face-swapping, but deepfakes have caught public attention in a way that other AI techniques have never really been able to. There is something very visceral about seeing a face where it doesn't belong, seeing an actor in a role you know that they didn't play, or seeing your own face doing something you've never done.

While the techniques that make up deepfakes have all existed previously on their own, together, they provide completely new possibilities. There are numerous use cases that deepfakes can be applied to, from stunt-double replacement to advertising. The technology is here, and its use will only grow as more and more industries find ways to use it.

There are still limits to the capabilities of generative AI. Knowing what a deepfake cannot do is as important as knowing what it can do. Especially regarding data, knowing how to work around those limitations is key to a quality result.

We've given an overview of deepfakes, covering what they are, what they can be used for, how they work, their limitations, and the existing software you can use to make them. In the next chapter, we'll cover the potential dangers of deepfakes and talk about the ethical questions that the technology brings with it.

2
Examining Deepfake Ethics and Dangers

Deepfakes have a reputation for being extremely dangerous and of questionable ethics. While any technology can be abused, it is the usage itself that is unethical, not the technology itself. This applies to deepfakes as well. As long as the usage is ethical, deepfakes can be an ethical technology. However, there are still potential threats from deepfakes. The technology is available publicly – anyone can make a deepfake, even those with unethical intentions. The best way to limit the danger of deepfakes may well be to make sure that everyone knows about them, effectively inoculating the public against trusting a video just because they have watched it.

In this chapter, we will cover the ethics and dangers of deepfakes. We will examine the ethics of deepfake's origins, create some guidelines to follow in creating ethical deepfakes, discuss the dangers of deepfakes, and cover some of the defenses against deepfakes.

We'll cover the following topics in this chapter:

- The unethical origin of deepfakes
- Being an ethical deepfaker
- The dangers of deepfakes
- Preventing damage from deepfakes

The unethical origin of deepfakes

Deepfakes began with an unfortunately unethical beginning. The first public deepfakes were created by a Reddit user by the name of /u/Deepfakes. These first videos showing off Deepfake's capabilities involved putting (usually) female actors into pornographic scenes. This non-consensual pornography was radically different from previous fakes in that they were more realistic, easier to create, and full video, capable of poses and expressions that the target actor had never performed in any recorded media.

Non-consensual pornographic deepfakes are not victimless events. This usage is extremely damaging as it harms the actors in profound ways. With non-consensual deepfakes being made, celebrities are losing control of their image and reputation. Unfortunately, there is no way to completely avoid the threat from deepfakes as even after proving a video isn't actually of them, the deepfake will still be spread and objectify the actor. Some of the people being deepfaked in pornographic films against their will aren't even celebrities and aren't equipped or ready to deal with the attention and infamy of the situation.

> **Author's note**
> While we believe our stance is clear, we want to once again reiterate that the authors do not condone or participate in non-consensual pornographic deepfakes. We believe that the practice is horrible and causes serious harm to its victims.

Putting women's faces into pornographic videos that they did not participate in or consent to is obviously a highly unethical use of any technology. Unfortunately, this is still the most well-known use of deepfake technology. However, this is slowly changing, with deepfakes becoming more recognized for their flexibility and innovation.

Being an ethical deepfaker

Deepfakes have many potential uses, both ethical and unethical. Knowing what is ethical or not is sometimes difficult to figure out. For example, are all pornographic deepfakes unethical? What if everyone involved consents? This can be difficult to answer sometimes, and there is a lot of ambiguity, but there are some guidelines that you can follow and some tests you can run to try to ensure your usage is on the right side of the ethical questions.

Consent

Probably the most important ethical concern is one of consent. Consent is quite simple: is everyone involved willing to participate? This is most obviously an issue when it comes to pornographic deepfakes – if not everyone is a consenting participant, then it's unquestionably unethical.

However, this becomes less obvious when the video is satirical or an actor is put into a role that they didn't originally perform. Very few deepfake creators can get explicit permission from an actor, and even if the actors were all okay with the usage, contacting the actors to get permission would be difficult.

Some deepfakes, such as the Jennifer Lawrence-Buscemi one by birbfakes, have received lots of attention (and in the case of Steve Buscemi, even a response) without direct permission from the people involved. These uses were likely ethical despite the lack of consent, as they were done in parody and without violating other ethical concerns. This can be a difficult balance to keep, but consent needs to be considered, even if it's hypothetical.

Even if an actor or actress has passed away, it doesn't mean that you don't need permission from their estate or family. Disney had permission from Carrie Fisher and her family to use CGI techniques to replicate her character in Rogue One and other Star Wars productions.

One important note is that while you should get permission (if possible) from the person you're swapping in, you should also consider the person you're swapping onto. While the harm there is less obvious, it's still a potentially harmful action.

Respect

It's important to remember that the people you're swapping to and from are real people who have emotions and have worked to be where they are. You should show respect to the people you're swapping.

In the case of pornography, this is again a point that is violated. Actors have to make an active decision in every role they look at as to whether they're comfortable with the content. This includes whether they're comfortable with nudity or sexuality in their roles. Even though they may have done some similar scenes in the past, they may change their mind if the work, context, or even opinions change. When a person who has chosen not to be in sexual roles gets deepfaked into pornographic scenes, their decisions are dismissed and violated. Actors who have participated in sexual scenes in the past were able to make that decision in the context of the work; that does not mean that they aren't harmed when put into other scenes in different contexts or put into scenes that perhaps they would no longer choose to participate in.

In the case of deepfake Tom Cruise by Chris Ume, and similar works, they stay on the positive side of this rule by showing nothing outrageous or disrespectful. The things that deepfake Tom Cruise does are not outside the realm of what Tom Cruise himself might do. The key thing to consider is that the fakes do not harm his reputation or demean his character and simply provide a glimpse of how the deepfakers believe Tom Cruise would act in those situations.

Deception

While the previous guidelines were all about the people being swapped, this one is actually about the audience. Creating deepfakes while being obvious and deliberate about them being deepfakes means that potential dangers are reduced and helps to keep them on the side of ethical usage.

This doesn't mean that your work can't be used unethically. For example, if it's taken out of context and shared on social media, it could spread far and wide without the warning that it's a deepfake. To combat this, the warning that it's a deepfake may need to be more than a disclaimer at the beginning or description. Try to think of how it could be misused and present the warning in a way that matches the risk.

If the risk of it being misused is low, then no explicit warning may be needed, instead simply allowing the absurdity of the situation or the innocuous nature of the deepfake to disarm the risk. This, for example, is the case with the large number of videos that have the actor Nicolas Cage swapped into them. It's easy to see that Cage wasn't the original actor (especially when a deepfake swapped Nicolas Cage onto Amy Adams in *Man of Steel*) and, even when it's not, the danger is low of any real damage ensuing.

Another clear example is when your deepfake is used where special effects or visual effects are commonplace. If you're bringing a deceased actor into a movie where visual effects would be expected, there is little reason to explicitly label the content as a deepfake. The context alone is sufficient to make it clear that you cannot trust everything you see. The famous story of the radio drama of H.G. Wells's *War of the Worlds* causing panic in the street is unlikely to happen with a deepfake bringing an actor back to life for a movie they couldn't film.

Now that we know how we can judge the ethics of a deepfake, how about we put this into practice.

Putting it into practice

In this section, we'll look at several deepfakes and evaluate them using the metrics that we proposed above.

Deep Tom Cruise

Deep Tom Cruise is a project by Metaphysic, a company that creates deepfakes for VFX and movies. They made several videos that deepfaked Tom Cruise in various situations and environments. They posted these videos on TikTok at `https://www.tiktok.com/@deeptomcruise`:

- **Consent**: For consent, we look at whether they had permission from those involved. Deep Tom Cruise did not get permission from Mr. Cruise. However, Cruise is a public figure so getting explicit permission may have been impossible. They did get permission (and participation) from Miles Fisher who was the impersonator whose performance they deepfaked into Tom Cruise.

- **Respect**: Mr. Cruise was never shown doing anything inappropriate or even the significantly out of character. The videos respected Cruise's reputation and personality. However, they did show Cruise doing things in his personal life that he wouldn't normally share, such as playing golf or having lunch, but these were not hurtful in any way.

- **Deception**: It was always very clear that Deep Tom Cruise was not a genuine account of Mr. Cruise. Even the name of the account itself disclosed that it was a deepfake. This was accompanied by interviews and news stories detailing that it was a deepfake, as well as how it was done.

- **Analysis**: While they did not get explicit permission from Mr. Cruise, they did not do anything to harm his reputation and clearly disclosed that it was a deepfake. On the whole, this is an ethical use of deepfakes, but it would have been better if they had gotten Cruise's explicit permission before creating the project. Presumably, if Cruise were to object to the project, they would remove it from public view and minimize the damage, as leaving it up with an objection from Cruise would tip it over into unethical usage.

Dalí Lives

Dalí Lives is an art project that shows an interactive deepfake of Salvador Dalí at the Dalí museum. This project's stated goals are to let visitors virtually "meet" and learn about the artist directly from Dalí himself. You can see information about the exhibit at `https://thedali.org/exhibit/dali-lives/`:

- **Consent**: Salvador Dalí is deceased and has no children or surviving immediate family. Consent is next to impossible in this situation.

- **Respect**: It's hard to imagine a more respectful usage than Dalí Lives. The museum worked hard to create a representation of Dalí so that people who visit the museum can interact with and see Dalí. The museum hosts many of his paintings and other artistic pieces and Dalí Lives fits into that environment perfectly.

- **Deception**: The museum is very clear that Dalí Lives is an art piece and not the genuine article. The project is set up in many locations around the museum along with placards explaining that the Dalí represented is not the actual person, but a project to let people interact with the representation. They also include a "making of" video showing the deepfake process.

- **Analysis**: This is pretty unquestionably an ethical use of deepfake technology. It's hard to imagine a more ethical usage than enabling visitors to a museum to interact with a famous historical figure who is no longer alive.

  While they lack consent, it's not hard to imagine that Dalí would have eagerly granted it if he were able. He loved the absurd and macabre. The idea that he could interact with guests from beyond the grave seems perfectly in line with his sensibilities.

  In addition, the project is respectful of Dalí, showing him in an environment that he fits perfectly into – a museum that has many of his own projects.

  Finally, there is no deception. Nobody who visits believes that they are truly interacting with Dalí and so this path is also clear.

While there are ways to approach deepfakes ethically, what dangers do deepfakes present?

The dangers of deepfakes

The dangers of deepfakes go beyond just fooling some people. If others act on what they see in deepfakes, there are huge dangers to reputation, politics, and even economics. Imagine if a large number of people were convinced by a deepfake of a declaration of war. There could be pandemonium and huge amounts of damage to international reputation before the truth was revealed. This very nearly happened recently: a deepfake of President Zelenskyy of Ukraine was spread online telling the Ukranian people to lay down arms during the Russian invasion.

Reputation

In some ways, danger to reputation sounds like the most trivial of potential dangers from deepfakes. However, reputation can be incredibly important for all kinds of things. Imagine if a person were to lose their job over a deepfake insulting their boss, or another person getting arrested when a deepfake falsely confessing to a crime is sent to the police. These effects can compound and seriously impair a person's future due to actions that they have no control over.

Wide-spread reputation damage is already incredibly easy in the internet age. People getting "canceled" for actions that they genuinely took can cause long-term harm to a person's future. Even if the damage is only temporary or limited, having the threat of actions you didn't actually do causing similar harm is a scary proposition.

Politics

The dangers of deepfakes in politics are much the same as the dangers to reputation but magnified by the public policy effects that can come from reputation damage. A politician, in many ways, relies on their reputation in order to get elected. Few would be willing to elect someone that they believed to be dishonest or disreputable (which is not the same as actually *being* dishonest or disreputable).

If a political group were to be targeted by a deepfake smear campaign, there are many who would believe the results even if they were revealed to be a fabrication. This is the unfortunate nature of human psychology where we are far more likely to believe something that confirms our already-held beliefs than something that disproves them. Once an opinion is made of a candidate in the minds of the public, there is little that can be done to reverse that opinion, especially bringing a reputation back from severe damage.

Deepfakes have, unfortunately, already been used in politics and have done varying amounts of damage. There are relatively innocuous uses such as Trey Parker and Matt Stone parodying then-President Donald Trump with Sassy Justice. Parody videos such as those are unlikely to do real harm as our society is already used to humor being used against politicians as a technique to criticize or entertain. In fact, many politicians actively participate in shows that parody them. One example is when Barrack Obama invited Keegan Michael Key to be his "anger translator" at the 2015 White House Correspondents' dinner – a role that originated in parodies on Comedy Central's *Key & Peele*.

However, we've already seen significant damage done to candidates over doctored videos. One video that spread among social networks in early 2019 was of Nancy Pelosi where the video had been slowed down to imply that she was inebriated or unfit as a politician. In Iraq, multiple female candidates withdrew from elections after pornographic videos claiming to be them were released publicly, harming their reputations.

Avoiding consequences by claiming manipulation

While it's easy to see the damage that a deepfake can cause, it's also possible for deepfakes to create issues where there is no deepfake. Now that deepfakes and manipulated media have entered the public perception, it's possible that genuine videos showing inappropriate behavior will be labeled as deepfakes and ignored.

This is unlikely to be a serious threat for minor faux pas or inappropriate activities as it's always been possible to diminish or dismiss videos (or even photos or voice recordings) that show a person doing things that are controversial. For more major situations such as videos of crimes or serious offenses, it's unlikely that the police or courts will long be confused or dissuaded by claims of deepfakes.

However, there are a range of potential scandals that may have no other way of defense beyond convincing people that the video is fake. In those situations, it's possible that the person involved may claim that the video is a manipulation as a defense against the controversy.

While it's good to know the potential dangers, arguably, it's more important to know how to defend yourself from the dangers.

Preventing damage from deepfakes

There is no foolproof way to make yourself immune to deepfakes or their effects, but there are activities and actions you can take in order to reduce the dangers. In this section, we'll examine methods that you can use to prevent as much damage as possible.

Starving the model of data

While deepfakes have been done with a minimum of data, the result is not seamless or high quality. This means that if you can simply prevent the deepfaker from getting enough data, you can prevent a deepfake from being of sufficient quality to fool anyone.

If you're trying to protect a famous person, this tactic is much harder, since every movie, TV show, interview, or even photoshoot that is available is a source of training data for a model. In fact, it's almost inevitable that enough data is available to target any famous person as the public nature of their job means that a lot of training data will be publicly available.

As a private individual, however, it is easier to control your media presence. By avoiding publicly available videos or images, you can prevent a deepfaker from getting enough content to create a convincing result. To do this, you can't allow publicly posted images or videos of yourself. This is harder with social media that allows your friends to tag you, allowing a deepfaker to potentially find media that you don't control being posted.

Authenticating any genuine media

There are services that now offer authentication of media. This lets users embed a verifiable watermark or metadata that allows the media to be proven to be genuine. Unfortunately, most sites and services remove metadata and recompress both images and videos, causing some types of watermarks to be removed. This means that even if you authenticate all genuine media, there will be nearly identical versions without the authentication embedded.

Authenticating media is an interesting solution to the problem of deepfakes, but may not be feasible: even if everyone starts supporting authenticated media, users will need to be trained to verify the authenticity, and that won't be possible until all major sites support media authentication. This leads to a chicken and egg problem, where media sites won't support authentication in the media until the public demands it and the public won't be aware of it until it's supported by all major sites.

Despite the failings of authenticated media, it might be a good idea to get a head-start on authenticating your media so that you'll be ready in the future when authenticated media is expected.

Deepfake detection

Another solution that relies heavily on large-scale support is **deepfake detection**. The idea behind deepfake detection is that a site would run any hosted videos through an algorithm or process to identify any deepfakes and either mark them as fake or block them from the site. In this situation, any site that had a perfect deepfake detector and a policy of flagging them would be safe from damage due to deepfakes.

Unfortunately, the solution is not that easy. First, deepfake detection requires dedication to detecting and flagging any deepfakes. Companies have paid lip service to detection, but ultimately, deepfake detection is not in use by any major services at this time. There are few economic incentives right now for companies to provide a thorough deepfake detection solution, but it's very possible that legal changes or public demand for deepfake detection could change the policy of companies so that they provide this detection.

> **A disclaimer on detection**
>
> At the time of writing this book, the authors are not aware of any social media or video content services that run deepfake detection on all hosted videos; however, it's possible that services are using detection to tag videos that the authors are not aware of.

Even if sites started dedicating themselves to detecting deepfakes, it's unlikely that they'd block them outright. There are legitimate uses of deepfakes and it's far more likely that sites would prefer to tag them (either publicly or internally) and monitor them than block these videos entirely. This would limit the effectiveness of any mitigations to the most flagrant violations, allowing lesser viewed or less offensive deepfakes to remain on the platform.

Another problem is in the accuracy of deepfake detection. There is no perfect way to identify and prevent deepfakes at this time. There have been many attempts to identify and flag deepfakes with many different solutions, ranging from purely technical to a mix of automatic and manual reviews. These methods can catch some but not all deepfakes right now. Even if they caught all the current deepfakes, the technology is moving and advancing – some developers will simply use deepfake detection as a way to improve their deepfakes to be undetectable again. This will lead to the same type of cat-and-mouse game that malware detectors have gone through. It's possible that deepfake producers may find new ways to fool detectors, even if just for a short time.

Finally, even if you have a perfect method for the detection of deepfakes, that would be of limited value in the long term. Even if one site started blocking deepfakes, there are a lot of other services that a deepfake could be posted on. Even if every major site blocked deepfakes entirely, there would be sites that would pop up specifically to deceive people with deepfakes. A total block of deepfakes online is impossible without complete control of the entire internet.

Public relations

If you have the money and the need, there have always been companies focused on helping to manage your reputation. They can be surprisingly effective and may be able to actively prevent deepfakes from being posted in the first place.

Public relations companies will have contacts with major companies such as Facebook and Youtube. They'll be able to get inappropriate or rule-violating images or videos removed much faster than attempting to do it manually. They'll also have legal departments and tools that may be able to get deepfake videos removed from other sources.

Even if a video can't be removed (or after it's already begun to spread), a public relations company could handle the fallout from the video, getting ahead of it, convincing news agencies to ignore the video, or pushing for stories about the video being manipulated.

The big issue is that public relations companies are not cheap and are selective about who they are willing to work with. Unless you are famous or have a very valuable reputation to maintain, it may be hard or prohibitively expensive to get a public relations company to work with you.

Public awareness

One other way to defend against deepfakes is to raise public awareness of deepfakes. Both the threats and potential, once publicly known, may encourage additional scrutiny of videos and images. This has the advantage of working even with new technologies or techniques but relies on public awareness and a willingness to question content.

Summary

The ethics and dangers of deepfakes are probably the most publicized aspects of them. However, they're things that all deepfake creators should consider, as it's easy to operate unethically if you don't consider these questions. Even those who aren't creating deepfakes may need to think about how to prevent damage from deepfakes or other manipulated media.

As a creator, you must consider the ethics of the content you're creating. Making sure that you evaluate such things as consent, the personal experience, and the labeling of a deepfake is a good way to clear ethical hurdles and make low-risk deepfakes.

Even as non-creators, people should be wary of the threats of deepfakes. The technology is "out of the bag" and cannot be erased. Even if the technology were to be banned immediately, state-level actors would still be able to wield it in potentially dangerous ways and could ignore any laws or restrictions placed on its use. Everything from reputation damage to political manipulation, to those who claim genuine videos are deepfakes to avoid repercussions are new threats that we can only mitigate through critical thinking and not through ignoring the subject entirely.

> **Author's note**
>
> The authors of this book are also developers of the open source deepfake software Faceswap. We have considered these same questions and the ethics of publicizing deepfakes. We consider the damage to be already done, and removing access to the technology now would not prevent any future abuses. We believe that the best way forward is to ensure that the public is aware of the threat so that critical thinking can be applied to any videos to consider whether they appear manipulated or out of the norm.

The dangers of deepfakes do have limits, and there are ways that they can be defended against. Right now, deepfakes require large amounts of data, and those who don't live in the public eye might be able to starve a deepfake creator of the content that they need in order to make a deepfake. For those who can't just remove all content, they can advocate for deepfake detection, media watermarking, or using public relations firms to help avoid damage from manipulated media.

In the next chapter, we'll get into the importance of data, how you can get enough data, and how you can make the most of the data you do have in order to get the best quality deepfakes possible.

3
Acquiring and Processing Data

Data is the most important part of any AI task, and deepfakes are no exception. The quality of a swap is limited by the input data, and selecting that data is the most important task of a deepfake creator. While there are ways to automate parts of data gathering, the process is still largely manual.

In this chapter, we will cover the importance of data, what makes quality data, how to get your data, and how to improve poor data. These are skills that are critical for a deepfake creator and must be developed. This chapter will explain the basics of these skills, but they must be practiced for full understanding and use.

Upscaling

We will cover the following sections:

- Why data is important
- Understanding the value of variety
- Sourcing data
- Improving your data
- Upscaling

Why data is important

Neural networks work by taking data that is known and processing it in order to train the deepfake AI (see *Chapter 1, Surveying Deepfakes*, for an explanation of the whole process). We call this set of data, simply enough, a **dataset**. To create a dataset, the data has to be processed and prepared for the neural network so that it has something to train with. In the case of deepfakes, we use faces, which need to be detected, aligned, and cleaned in order to create an effective dataset.

Without a properly formatted and prepared dataset, the neural network simply cannot be trained. There is another potential problem when it comes to generative networks like deepfakes – a poor quality dataset leads to poor swaps. Unfortunately, it's hard to know at the beginning whether a dataset will produce a good swap or not. This is a skill that takes time to learn, and your first few deepfakes are unlikely to turn out well as you learn the importance of data.

Time spent cleaning and managing data is time very well spent. While the largest time sink in a deepfake is the time spent training, that time requires no input from the creator – however, if your data is flawed, that time will be entirely wasted. To this end, most deepfakers spend a good deal of time cleaning and testing their data before they commit to a long training session.

You may think that **resolution** (the number of pixels in an image) is important in your dataset, but it's really not that important. There are very few high-resolution deepfakes available. A video may be 4k, 8k, or even higher, but the face swapped will be limited by the AI model, the capabilities of the computer it's trained on, and the time available to spend training. 512x512 pixels is a very high resolution in the world of deepfakes, and higher resolutions are reserved for the most dedicated and skilled deepfakers.

That doesn't mean you can't get good results with lower resolutions. One clear way to think of the difference is that **fidelity** is not the same as resolution. Fidelity measures how little information is lost from its original, that is how realistic or lifelike an image is. Think of a small thumbnail picture of a realistic painting, such as the image of the Mona Lisa shown in *Figure 3.1*: it is *highfidelity* but *low-resolution*. The opposite in this example would be a drawing by a 5-year-old (or in this case, one of the authors) of the Mona Lisa using an advanced drawing pad on a computer. That image is *high-resolution* but *low-fidelity*. As you train, your deepfake AI will increase its fidelity but the resolution will stay the same.

Figure 3.1 – "The Mona Lisa" by Leonardo DaVinci, a high-fidelity but low-resolution image
(UL); "A Mona Lisa" by Bryan Lyon, a high-resolution but low-fidelity image (LR)

Data is very important to your deepfakes, and variety is the most effective way to ensure good data.

Understanding the value of variety

Variety is the single most defining trait of a good dataset. The best datasets will all have a large variety of poses, expressions, and lighting situations, while the worst results will come from data lacking variety in one or more of these categories. Some of the areas of variety we'll cover in this section include pose, expression, and lighting.

Pose

Pose is a simple category to both see and understand. Pose is simply the direction and placement of the face in the image. When it comes to deepfakes, only the pose of the face itself matters – the rest of the body's pose is ignored. Pose in deepfakes is important so that the AI can learn all the angles and directions of the face. Without sufficient pose data, the AI will struggle to match the direction of the face and you'll end up with poor results.

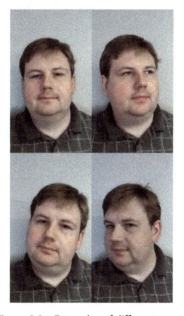

Figure 3.2 – Examples of different poses

One quick way to get the full range of poses is to simply move the head while being filmed. By looking around up, down, and from side to side (along with mixtures of these), you give the AI all the directions that it needs. Don't forget to get distant shots with a high **focal length** to zoom into the face, as well as close-ups with a neutral focal length. A high focal length "zoom" on a distant face causes the "flattening" (the effect is properly called **lens compression**) of the depth of the face and is often used for dramatic effect in movies, so you need to match the variance in your data as well. Even something such as lens distortion matters and can affect the image enough that getting a variety of examples can help get quality data.

Expression

Expression is another aspect that is easy to understand – it's the shape the face makes during training. Unfortunately, it's very hard to ensure that you have met the full range without missing important data. Think about how expressive a face may be – happy, sad, angry, awake, surprised, and so much more – and think about how easy it may be to miss an important subset of those expressions in your data.

There is no easy way to ensure that you have all the variety that you need, but the more the better. You don't want to just sit someone down and film their face for an hour, as you are unlikely to get all the expressions you really want. To really get a good variety of expressions, you need to make sure that you get data from different days, in different environments, and with different emotional stimuli. An actor may give you drastically different "sad" expressions in different contexts or times, and that variety is important for the AI to really understand how to recreate those same expressions.

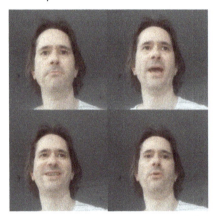

Figure 3.3 – Examples of different expressions

The key here is to get as many different expressions as you can. You can't get all your data at once; if you do, your data will be extremely limited. If you have an excellent actor as your target face, you might be able to get away with having them match the expressions that you're swapping to, but you'll probably want to get a much more varied dataset to ensure you're getting the right data in the right contexts.

Lighting

Lighting may sound easy and/or trivial but ask any photographer or videographer about it and they'll be able to talk forever. Lighting a photo or video shoot is complicated even when you're not dealing with AI, but when deepfakes are involved, it becomes an enormous issue. It's simply impossible to get a complete variety of lighting, but there are some keys to consider.

"Good" lighting

Most Hollywood films are examples of expert light management with entire teams dedicated to even the simplest setups. However, there are a few good ground rules for basic lighting for deepfakes.

Clear lighting cannot come from a single direction. This leads to deep shadows in the eyes, next to the nose, and often on one whole side of the face. Even direct face-first lighting will lead to shadows at the edges (not to mention probably lead to expression changes as the actor squints). Instead, you need to have good **ambient**-style lighting – that is, lighting coming evenly from all directions. Offices with large fluorescent lights will probably be a good example of this lighting (although the color of the light will leave something to be desired).

Shadows

However, ambient lighting is not the only style of lighting that deepfakes require. You need some shadow in the dataset or the AI will never learn how to recreate anything but a fully lit face. For this lighting, you can use a normal overhead light or lamp. This will enable the AI to pick up some of those deeper shadows that really help to bring a face's depth into perspective.

The best way to get shadow variety is in scenes where the lighting or subject is moving, giving you lots of varied shadows that the AI can learn.

Outdoors

Finally, you are going to want to get some outdoor images. The Sun is particularly unique among all other light sources. There is nothing that quite matches the natural effect of a burning ball of plasma millions of miles away from the subject – and believe it or not, the distance really does matter, as it leads to basically parallel light rays. Because of this, no matter how much "in-studio" filming you do, it's very important to get at least some of your data outdoors. The Sun's light is truly impossible to reproduce, and getting data that was filmed outdoors is indispensable to fill out the lighting data for the AI.

Bringing this variety together

It's important to think about all the aspects of variety separately, but you actually also need to consider them together as well. You should get data that mix and match all three aspects in order to ensure that you get the best results. For example, all your different lighting scenarios should have a variety of expressions and poses. Only with varied data can the AI learn to create a truly convincing deepfake.

Having good variety is critical, but how do you actually get the data you need?

Sourcing data

If you're working on a deepfake, then you probably already know who you're going to be swapping. Hopefully, you're lucky and are working on a deepfake that covers two people who you have access to so you can film them or gather data from them without too much trouble. Unfortunately, not everyone is so lucky, and most of the time, one or both of your subjects will be unavailable for custom data (this is probably why you're working on a deepfake in the first place, after all).

These two situations require very different approaches to getting data, and sometimes, even if you have good access to your subjects, you will need to get some of your data from another source.

Filming your own data

Filming your own sources is a dream position for any deepfaker. The ability to build the perfect dataset by putting an actor in front of a camera is liberating, but also a wasted opportunity if you don't know how to capture all that you need.

> **Author's note**
>
> This book's scope is unable to even begin to approach the extensive vocabulary and techniques that photography and filming involve. If you're planning on doing much actual filming of your own data, it is highly recommended that you read some photography or filmography books, watch some tutorials, and talk to experts in that field before you get on set so that you can get the most out of the time.

Telling the **gaffer** (the person in charge of lighting on a set) to go against all their training and deliberately give you bad lighting can be scary if you're not an expert. Telling an actor to make a "sadder" face or to make extreme expressions can be equally awkward. When filming for a deepfake, you need to make it clear to the crew that you do need their regular skills, but that you also need to go beyond the norm into the weird, esoteric, and uncomfortable.

While you may be tempted to ensure that the data is always done in close with the actor's face filling the frame, don't be afraid to let the face be smaller or farther from the camera; it's rarely the resolution of the face in the data that hurts the results.

You need to consider all aspects of a deepfake when you're filming. Even when filming the video to be swapped, it is important to consider the limitations of the deepfake technology. Angled faces, such as profiles, are especially difficult to swap without an uncanny effect. Because of this, you will want to minimize the number of faces looking far from the axis of the camera. They don't have to be directly facing the camera, especially if you get a lot of angles in the training data, but there comes a a point when things things are just too angular for the AI to provide a good result.

One important detail is that your deepfakes will improve with practice as you develop your skills. For this reason, it is highly recommended that you not rely on your first deepfake to be perfect, and if you have an important project coming up that requires deepfakes, you may want to create a couple of practice deepfakes first so that you can get the necessary practice with the data to ensure that while filming your important deepfake you have the experience necessary to get quality data.

Getting data from historical sources

Sometimes, you cannot film your subjects – for example, if the actor has passed away or you're deepfaking a younger version of an actor. In these cases, you don't have the luxury of making your own data to meet your situation. That means you must rely on older sources for all your training data.

Hopefully, your subject is recent, and videos and photos of them exist. Some subjects cannot be deepfaked due to the sheer lack of original content around them – Albert Einstein, for example, will never have a traditional deepfake made simply because there isn't the necessary training data. Your best bet at that point is to use an impersonator to some degree. This is the technique that the Dalí Museum used to bring Dalí back – impersonators and some limited video interviews of the artist.

If you're lucky enough that data exists for your subject, you can use any data available, and you do not need to restrict yourself based on age or resolution (see the discussion about the difference between resolution and fidelity in the *Why data is important* section). Instead, focus more on getting variety. It can be a challenge knowing where to stop but, generally, time spent gathering and managing your data is well rewarded. It's unlikely that time spent on data will be wasted in the long run.

Movies, interviews, and even random photos of people can be used for deepfake training data. You don't want to rely too much on any one of these, as they are all very limited in domain. A horror movie is unlikely to have many smiles and random photos that you can find will be unlikely to have anything besides smiles. Finding the balance between the different categories in order to find the variety necessary to train is one of the skills that you will simply have to learn.

Especially when dealing with historical data, you'll often find that it is not as high-quality as you want, so how can you get the most out of your limited data?

Improving your data

There are no silver bullets to magically make your data better, but there are some ways that you can tweak your data to improve the training of the AI.

Linear color

When you're filming you may film in **logarithmic scale (log)** color, where the scale represents an exponential change. This is great for storing a large color range while filming but does not work well for training a deepfake. To get the best results from your training, you'll want to convert your video into a **linear** color space (where a change of some number is consistently represented). It doesn't really matter which one, but all your data and converted videos should be the same. Since most content is **Rec.709**, we recommend that you use that unless you have a good reason to pick a different color space.

> **Author's note**
>
> Color science is a very robust field, and a full examination is outside the scope of this book. However, a basic understanding can help. **Rec.709** is one of many so-called **color spaces**, which are how a computer or TV represents colors. There are many different color spaces, including **SRGB** and **Adobe RGB** (among others), which can show different colors or represent them in different ways. The important thing when it comes to deepfakes is that you must keep all your data in the same color space when training and converting.

High-dynamic range (HDR) content is also problematic – at least when it comes to commercially-released content. Consumer HDR technologies throw out a lot of data to fit the color data into the frame. This means that something that looks dark in one scene could actually be stored as brighter than something that looks bright in another scene. Even when mapped back to a linear range, it tends to be highly variable, which causes great difficulty for AI to generalize, often leading to the failure of the deepfake model. The best way to deal with this is to avoid giving HDR content to the AI model. This can be complicated since sometimes the highest quality data you can get is HDR. In that situation, it's useful to know that data of a lower resolution but a consistent color space will lead to better results than a higher resolution but inconsistent color.

Professional cameras recording in RAW will not have the same problem as **HDR** since they keep the exact data coming from the sensor and the complete color data remains available.

Color science is complicated and sometimes non-intuitive. It's not necessary to become an expert in all the details, but it's useful to know when the color space might be causing a problem.

Data matching

Sometimes, you're extremely limited with the data that you have access to for one of your subjects. In that case, it can be very hard to get a good swap. One way to help get the results you need is to match your data as close as possible for at least part of your training. Have the subject that you have access to match the data for the other subject as closely as possible. It's important to match the poses, expressions, and lighting all in a single image. This doesn't have to be for your final conversion and shouldn't be the only data you use, but having closely matching data can sometimes help the AI find the details to swap the two.

When doing data matching, you'll want to train for a very short time on just the subset of your data which matches. Both your subject datasets should be cleaned to a single short segment and training should be limited to no more than 100 **epochs** (an epoch is one full cycle of training the AI model on all the data). Once you're done training with that subset, you'll want to train with the full data. You may want to repeat this process a couple of times with different subsets of your data.

Upscaling

One major current trend in AI is in **upscaling** content. This is a great use for AI, which can fill in missing data from its training, especially **temporally aware upscalers**, which can find missing data by tracking an object across multiple frames to get more detail. Unfortunately, when used as training data for generative AI such as deepfakes, the AI upscaled data is problematic and prone to training failures. Even a very good upscaling AI has glitches and artifacts. The artifacts might be difficult for the eye to see, but the deepfake AI searches for patterns and will often get tripped up by artifacts, causing the training to fail.

Generally, the best way to deal with upscaling is to not upscale your training data but instead to upscale the output. This is even better in many ways since it can replace missing face data and improve the resolution of the output at the same time. The reason that chaining AIs in that direction doesn't cause failures is that, unlike deepfakes, upscalers are not trained on the data generated.

That said, there are a lot of techniques that are completely safe for upscaling your training data. More traditional or temporal solutions that don't involve AI can be used to upscale without leading to failure to train. Upscaling with **bicubic** or **Lanczos** filtering is also perfectly acceptable.

Cleaning up noise or adjusting colors is also fine and a valid way to maximize your data quality, especially if the extract was unable to find faces in some of the data.

Summary

Data is critical for deepfakes, as with all AI. Getting the best data is a skill that you have to learn over time. It's something that you will get better at as you become more experienced with deepfakes. That being said, some tasks can be learned without heavy investment in the process. Cleaning and organizing your data is important – time spent on this can save you time later since your AIs will be less likely to fail.

Filming your own data is sometimes necessary and can get you the best results, as this will give you enough control to fill in missing data or match limited historical data. When you have nothing but historical data, you're more limited and may need to do further work to improve the data you have. Upscaling and filtering are possible, but you must be careful, as some techniques can add artifacts that interfere with training.

In the end, data is the most important part of training a deepfake and therefore is the most important job of a deepfaker. You must spend time and effort learning the skill of data management if you are going to excel at creating deepfakes.

In the next chapter, we will be walking you through using Faceswap – a freely available open source deepfake software – so that you can generate your own deepfakes.

4

The Deepfake Workflow

Creating a deepfake is an involved process. The tools within the various software applications help to significantly reduce the amount of manual work required; however, they do not eliminate this requirement entirely. Most of this manual work involves collecting and curating source material, as well as cleaning up data for the final swap.

Whilst there are various applications available for creating deepfakes this chapter will use the open source software Faceswap (`http://www.Faceswap.dev`). The general workflow for creating a deepfake is the same from application to application, but you will find the nuances and available options vary between packages.

It is also worth noting that Faceswap, at its core, is a command-line application. However, it also comes with a GUI that acts as a wrapper to launch the various processes. Within this chapter, the GUI will be used to illustrate the workflow; however, most of the tasks performed here can also be run from the command-line.

In this chapter, the deepfake workflow will be covered from the inception of the swap to the final product. Specifically, the following topics will be covered:

- Identifying suitable candidates for a swap
- Preparing the training images
- Training a model
- Applying a trained model to perform a swap

Technical requirements

As with all machine learning techniques, deepfakes can be created on any PC with a minimum of 4 GB of RAM. However, a machine with 8 GB of RAM or higher and a GPU (a graphics card) is strongly recommended. Training a model on a CPU is likely to take months to complete, which does not make it a realistic endeavor. Graphics cards are built specifically to perform matrix calculations, which makes them ideal for machine learning tasks.

Faceswap will run on Linux, Windows, and Intel-based macOS systems. At a minimum, Faceswap should be run on a system with 4 GB of VRAM (GPU memory). Ideally, an NVIDIA GPU should be used, as AMD GPUs are not as fully featured as their Nvidia counterparts and run considerably slower. Some features that are available for NVIDIA users are not available for AMD users, due to NVIDIA's proprietary CUDA library being accepted as an industry standard for machine learning. GPUs with more VRAM will be able to run more of the larger Faceswap models than smaller GPUs.

It is also possible to rent cloud services (such as Google's Cloud Compute or Amazon's AWS) to run Faceswap remotely.

Installing and setting up the software will not be covered in this chapter, as the method will vary between OSs, and detailed installation instructions can be found on the Faceswap website (`https://forum.Faceswap.dev/viewforum.php?f=4`).

Identifying suitable candidates for a swap

While it is technically possible to swap any face with another, creating a convincing deepfake requires paying some attention to the attributes of your source and destination faces. Depending on what you hope to achieve from your deepfake, this may be more or less important to you, but assuming that you wish to create a convincing swap, you should pay attention to the following attributes.

- **Face/head shape**: Are the shapes of the faces similar to one another? If one face is quite narrow and the other quite round, then while the facial features will be correct, the final swap is unlikely to be particularly convincing if the final swap contains a head shape that is significantly different from the individual you are attempting to target.

- **Hairline/hairstyles**: While it is possible to do full head swaps, these are generally harder to pull off, as hair is complex, and hairstyles can change significantly. You will generally be swapping the face but keeping the hair from the original material, so you need to keep hairline and hairstyles in mind when considering your swap.

- **Skin tone**: The neural network will do some work in matching the skin tone between the faces that you train the model on; however, this will only work to a certain extent. As some attributes of the original face are likely to still exist within the final swap when it is blended into the original frame, it is important to ensure that the natural skin tone between the faces is not significantly different.

Once you have identified your candidates for creating a deepfake, it is time to collect data to train the model.

Preparing the training images

In this section, we will be collecting, extracting, and curating the images to train our model. Far and away the best sources for collecting face data are video files. Videos are just a series of still images, but as you can obtain 25 still images for every second of video in a standard 25 FPS file, they are a valuable and plentiful resource. Video is also likely to contain a lot more natural and varied poses than photographs, which tend to be posed and contain limited expressions.

Video sources should be of a high quality. The absolute best source of data is HD content encoded at a high bitrate. You should be wary of video content acquired from online streaming platforms, as these tend to be of a low bitrate, even if the resolution is high. For similar reasons, JPEG images can also be problematic. The neural network will learn to recreate what it sees, and this will include learning compression artifacts from low-bitrate/highly compressed sources. Footage filmed on a modern-day smartphone or better, or extracted from Blu-ray or DVD sources, is ideal. One caveat to this is that you should avoid using HDR footage at all costs. HDR, by its very nature, contains images within a dynamic range. Neural networks expect the data they receive to be within a consistent range, so they struggle with HDR data and, quite often, cannot learn at all when provided with this kind of data.

You are looking to collect material from as many different sources as possible. It is a misconception that a model is trained for a specific scene. Neural networks of the type that deepfakes use benefit from highly varied data. As the neural network is looking to encode important features for each of the faces it sees, giving it as much varied data as possible will enable it to generate better encodings and better feature maps. This includes variety in poses, expressions, and lighting conditions. While the neural network will do some work to simulate different lighting conditions, this is to help augment already varied data rather than act as a replacement for missing data. The neural network will not be able to create poses and expressions that are significantly different from anything it has seen before.

Extracting faces from your source data

Now that you have a variety of sources, the next step is to extract and **align** the faces (a process that normalizes the faces in the images) from these sources to build your training set. You are looking to collect between 500 to 50,000 faces for each side that you intend to train on. The variety of data is more important than the quantity of data. Five-hundred highly varied faces will lead to far superior results than 50,000 near-identical faces.

During the extraction process, an "alignments file" will be created. This file (with a `.fsa` extension) contains information about the faces that have been discovered within each of your sources. With this in mind, it is good practice to set up a project folder structure to store your data so that you can easily locate and edit any of the data as required. A reasonable structure could be along the following lines:

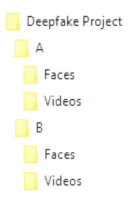

Figure 4.1 – A suggested Faceswap folder structure

For each side of the model (**A** and **B**), we are creating a folder to store the source videos in, along with their associated generated alignments files (**Videos**), and a faces folder to store the extracted faces (**Faces**). If you are extracting from images as a source for faces, then you can add an **Images** folder.

Copy the video and image source files to their associated folders and launch the Faceswap application:

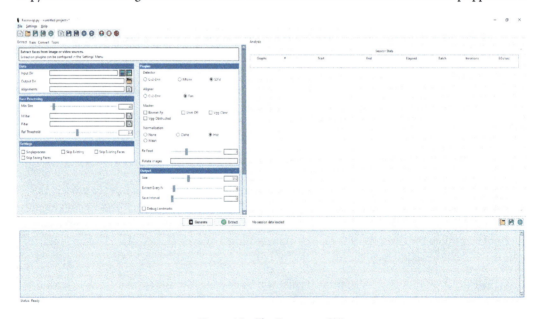

Figure 4.2 – The Faceswap GUI

The application is divided up into separate sections, depending on the task that you are currently performing. At this stage, we are extracting faces, so make sure that the **Extract** tab is selected.

There are many options available here, but we will just be focusing on those that are required to generate a training set. Optional or unnecessary options will be skipped for brevity, but you can view the tooltip for more information in the GUI.

> **Tip**
>
> Faceswap comes with built-in **tooltips** that explain what each option does. Hover over the entries to access the corresponding ToolTip.

Data

This section is where the location of the source material that we intend to extract faces from is entered, as well as the location that we wish to extract the faces to:

- **Input Dir**: The location of the source video file or folder of images that contain the faces you wish to extract. Clicking the buttons on the right will launch a file browser to easily navigate to the correct location. Select the left-hand icon if extracting faces from a video file or the right-hand icon if extracting faces from a folder of images.

- **Output Dir**: The location that identified faces should be extracted to. This should be a new folder within the **Faces** folder that you set up when creating your project folder structure.

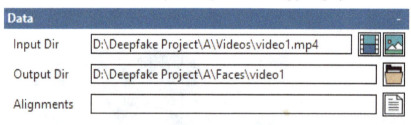

Figure 4.3 – The Data section for face extraction

Next, let's look at the plugins we can use.

Plugins

The plugins section is where the neural networks that will we use to identify faces within the images are chosen, as well as those plugins that identify the key landmarks within the face and any neural network-based masks to apply to the extracted faces:

- **Detector**: The detector identifies faces within each of the images. The most robust detector at the time of writing is **S3Fd** (based on the paper Single Shot Scale-Invariant Face Detector: https://arxiv.org/abs/1708.05237). It is, however, resource-intensive, requiring at least 3 GB of GPU memory to run. It also runs incredibly slowly on a CPU due to its complexity. However, if you have the resources available, this is the detector to use. Otherwise, the **MTCneural network** (https://arxiv.org/abs/1604.02878) detector should be used, which will run on far fewer resources and runs a lot quicker on a CPU.

- **Aligner**: Responsible for identifying key facial landmarks. These landmarks are used for aligning any faces detected so that they are consistent for feeding the model during training. **FAN** (`https://arxiv.org/pdf/1703.07332.pdf`) is the best aligner and, if at all possible, should be the option you select here. It is, however, slow on a CPU, in which case, the **CV2-D** neural network aligner is available. While this will run a lot quicker on a CPU, it is far inferior to FAN.

- **Masker**: When extracting faces, it is also possible to use neural networks for masking an area of interest. Specifically, we are only interested in training our model on the face area of the extracted image. The background to the face that is included in the extracted images just adds noise to the model, which it is advantageous to exclude. Two masks are always included by default, based on the landmark data generated by the aligner. The landmark-based masks are fine for a lot of use cases, but they are limited insofar as they do not include the forehead within the masked area (they crop just above the eyebrows). The neural network-based masks attempt to address this issue by using AI to generate masks on extracted faces. Generally, the **BiSeNet-FP** mask works best.

It is worth noting that masks do not need to be created at extract time. Faceswap includes a tool to add masks to training sets after extraction has been performed. This can be beneficial, as often you will not want to spend time generating neural network-based masks until you are happy that the extracted faces are correct and properly aligned. Be aware that the more neural network-based masks that are added (multiple masks can be selected), the longer the extraction process will take.

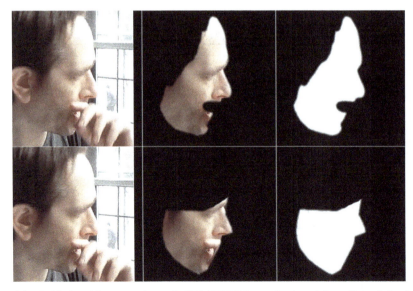

Figure 4.4 – A side-by-side comparison of a BiSeNET-FP (neural network-based) mask (top) and a components (landmarks-based) mask (bottom). The left image is the original aligned face, the right image is the generated mask, and the center image is the aligned face with the mask applied

- **Normalization**: When the aligner is looking to identify key landmarks, it can help to normalize the image being fed to the plugin. Generally, **histogram normalization (Hist)** or **contrast limited adaptive histogram equalization (CLAHE)** works best, but it will depend on the source material.

- **Re Feed/Rotate Images**: To generate a training set, these options are not necessary and should be left at their default values (**1** and blank respectively).

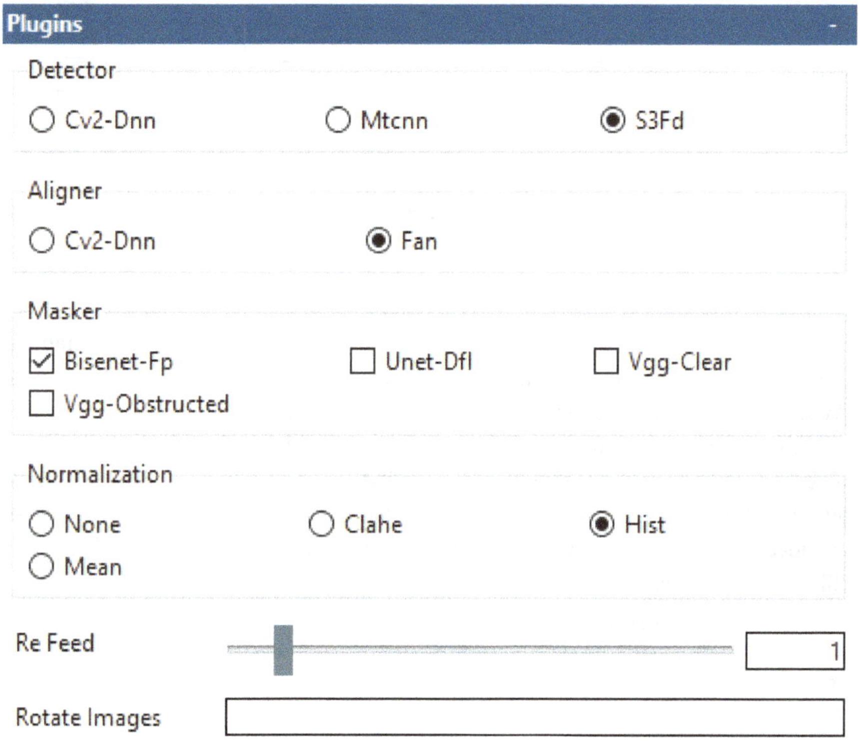

Figure 4.5 – The Plugins selection options for face extraction

Face processing

Face processing is any task that should be performed after faces have been identified within an image. The only option of concern here is **Min Size**. False positives can be found by the detector (items that the detector considers a face but are not actually so). This option allows you to discard faces that do not meet this minimum threshold (measured in pixels from corner to corner of the detected face). The value specified here will vary, depending on the size of the input frames and how many of the images are taken up by the faces you are interested in. Leaving it set at a low value, regardless, can help with the curation of data within the next phase.

The other option within this section is the face filter. Generally, it is advised to avoid using the filter, as it can be somewhat hit and miss at correctly identifying faces and will significantly slow down the extraction process.

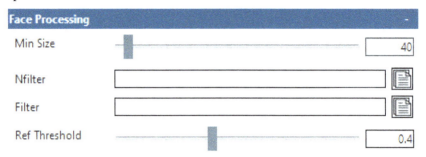

Figure 4.6 – The Face Processing options for face extraction

Output

The final area of interest is the **Output** section. The only option that needs to be amended from default is the **Extract Every N** option. While you need lots of images to train a model, variety is more important. When using video as a source of training images, each frame is extremely similar to its immediate neighbor. To reduce the number of similar faces that will be extracted, it is possible to skip frames to parse for faces. The number that is set here will depend on the frame rate of your video, as well as the number of sources that you intend to extract faces from. Generally, between 2 to 8 frames per second of video is a good number to aim for (for example, for a 30 fps video, an **Extract Every N** value of 6 will extract faces from 5 frames for every 1 second of video).

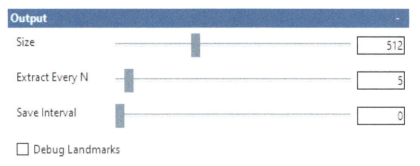

Figure 4.7 – The Output options for face extraction

Extract

Now that the options have all been set, the **Extract** button can be pressed. This will extract all of the discovered faces into your given folder and generate the corresponding alignments file. The amount of time this will take will depend on the available hardware, the length of the source material, and the plugins that have been chosen.

Curating training images

Once faces have been extracted, the data needs to be curated. The neural networks used to identify, extract, and align faces do a good job, but they are not perfect. Along with the correctly detected and aligned faces, it is most likely that a not insignificant amount of unusable data will also have been collected. This may include faces other than the target, false positives (parts of the image that the neural network considers a face but are not actually so), and misaligned faces (faces that have been correctly identified but the aligner has failed to align them correctly).

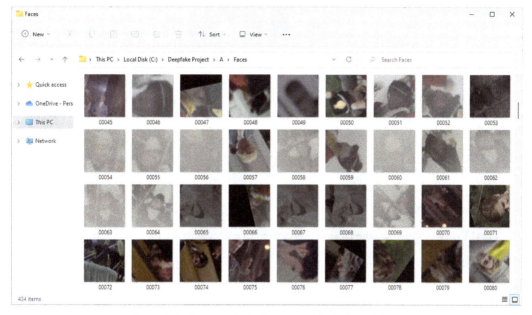

Figure 4.8 – A concentrated example of misaligned faces

Examining a folder that contains a significant number of unusable faces, it may, at first, appear to be a mammoth task to clean up and curate the training images. Fortunately, neural networks can again be leveraged to make this job a lot easier.

Sorting the faces

The faces will have been extracted in frame order – that is, they will exist within the output folder in the order that they were discovered within the source video or images. Faceswap includes a number of sorting mechanisms to arrange these faces in an order to enable easier pruning, which can be accessed by selecting the **Tools** tab, followed by the **Sort** sub-tab:

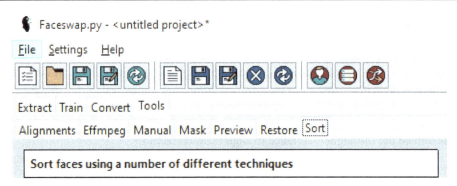

Figure 4.9 – The Sort tool within the Tools section of Faceswap

The most powerful sorting method, by some distance, is to sort by **face**. This uses the neural network **VGG Face 2** developed by researchers at the Visual Geometry Group at the University of Oxford (`https://arxiv.org/abs/1710.08092`). Faceswap utilizes this network to cluster similar faces together, making the data far easier to parse.

Within the **Sort** section of Faceswap, the following options should be selected:

- **Input**: The folder that contains the extracted faces that are to be curated
- **Sort By**: Select **Face**
- **Final Process**: Rename (the faces will be sorted in place, with the filenames renamed)

All other options can be left at their default values, as shown in the following screenshot:

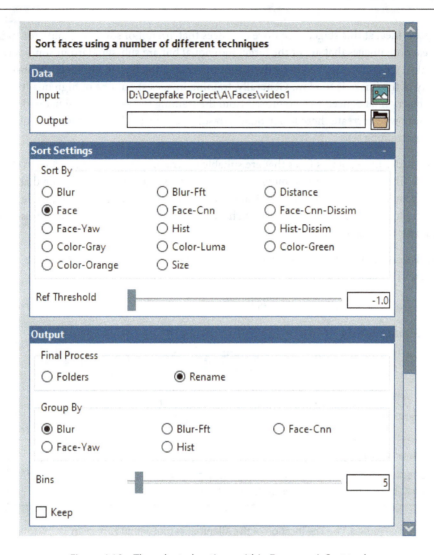

Figure 4.10 – The selected options within Faceswap's Sort tool

Press the **Sort** button to launch the sorting process. Depending on your setup and the number of faces to be sorted, this may take some time.

Removing faces

Once the sorting process has completed, the OS's standard file manager can be used to scroll through the folder and remove any incorrect faces. The sorting process will have grouped all of the similar faces together, making this a far simpler task, as all of the misidentified faces can be bulk selected and deleted. Just ensure that the sort order of the folder is by **filename**.

It is also a good idea, at this stage, to remove any faces that are not of suitable quality for training. This includes those images that are of the target individual but are sub-standard in some way – for instance, if the face is not aligned correctly within the frame, the face is significantly obstructed, or the quality of the image is too low. Generally, training images should be of high quality. Not all low-quality/blurry images need to be removed from the training set, as the neural network will also need to know how to recreate these lower-quality/resolution images; however, they should form a minority of the training set.

Now that the folder just contains faces that are suitable for training, it is a best practice to clean the alignments file. This is the process of removing the faces that you have just deleted from that file. This enables us to go back to the video source and alignments file and re-extract the faces, while avoiding the requirement to sort the data again. This action has a second advantage, insofar as it will also rename all the faces their original filenames.

Once again, Faceswap has a tool to help with this – specifically, navigate to the **Tools** tab and then the **Alignments** sub-tab:

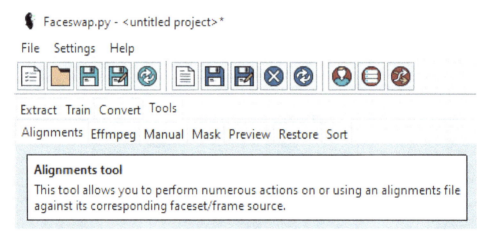

Figure 4.11 – The location of the Alignments tool within Faceswap

The **Alignments** tool allows multiple actions to be performed on the alignments file. The job that is of interest is **Remove-Faces**, which examines a folder of faces and removes the faces that you have deleted from the alignments file. The following options should be selected:

- **Job**: Select **Remove-Faces**

- **Alignments File**: Browse to the location of the source video and select the corresponding .fsa alignments file

- **Faces Folder**: Browse to the location that contains your curated face set and select the folder

All other options can be left at their default values:

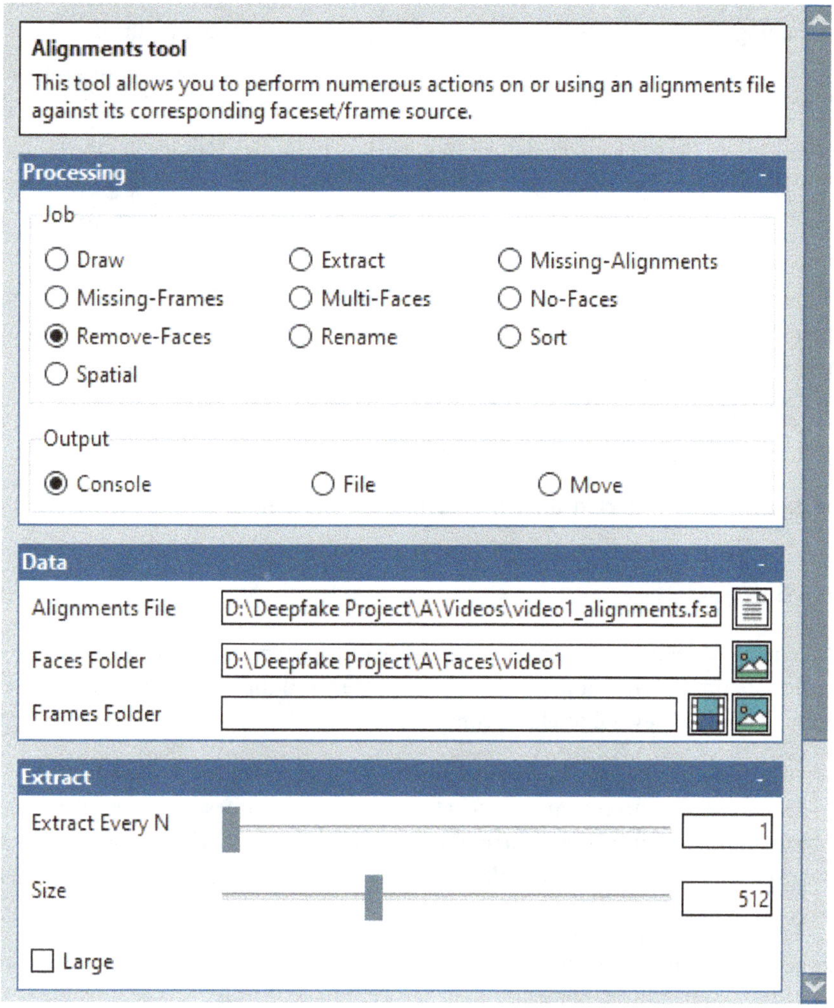

Figure 4.12 – The selected options within Faceswap's Alignments tool

Press the **Alignments** button. A backup of the alignments file will be taken, then the process will remove all the faces that do not appear in the faces folder from the file. Finally, the faces will be renamed with their originally extracted names.

Collating training images

Now that the training sources have been curated, the images can be collated into the final training sets.

> **Tip**
>
> If the required mask was not generated during extraction, it can also be added to the faces now, using the **Mask** tool (the **Tools** tab | the **Mask** sub-tab). This should be done prior to collating the final training sets.

This is as simple as taking all of the contents of each source's extracted faces and placing them all into the same folder (one folder for the **A** side, and one folder for the **B** side). All the information required by Faceswap to train on these images is stored within the EXIF header of the PNG images.

Some final curating may be required to bring the number of training images down to a manageable size, but anywhere in the region of 500 to 50,000 images per side of the model is reasonable.

Now that the data has been collected and curated, it is time to train the model.

Training a model

This part of the process requires the least amount of manual intervention but will take the longest in terms of compute time. Depending on the model chosen and the hardware in use, this can take anywhere from 12 hours to several weeks to complete.

It is advised to use a relatively lightweight model when creating a deepfake for the first time. Creating swaps is fairly nuanced, and understanding what works and what doesn't comes with experience. Whilst Faceswap offers several models, starting with the **Original** or **Lightweight** model will allow you to gauge the performance of the swap relatively quickly, while not necessarily giving you the best possible final result.

Faceswap comes with numerous configurable settings for models and training. These are available within the **Settings** menu of the application. To cover all of these settings is well outside of the scope of this walk-through, so default settings will be used unless otherwise stated.

Setting up

Navigate to the **Train** tab of the Faceswap application:

Figure 4.13 – The training options section of Faceswap

Faces

This section is where we tell the process where our training images are stored. If you followed the extraction and curation steps, then these faces will exist within your project folder, with a single folder for each side:

- **Input A**: The path to the folder containing the extracted faces for the **A** side of the model (that is, the original face that is to be removed from the final video).

- **Input B**: The path to the folder containing the extracted faces for the **B** side of the model (that is, the face that you wish to transpose to the final video).

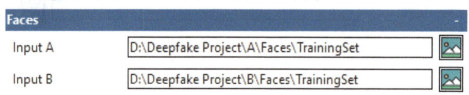

Figure 4.14 – The Faces options for training a model

Model

The model section is where Faceswap is instructed on which model to use, where that model should be stored, and any model-specific actions to perform at runtime:

- **Model Dir**: The folder that the model should be stored in, or if you are resuming a pre-existing model, then the folder that contains the model to be resumed.

  If you are starting a new model, then this location should not pre-exist on the hard drive. When the model is created, the folder specified here will be created and populated with the model and associated files required by Faceswap to track training.

  If you are resuming a previously created model, then this should point to the folder that was created when initially setting up the model (the folder created by Faceswap containing the associated model files).

- **Trainer**: Faceswap has multiple models available (named **Trainer** for legacy reasons). These models are more or less configurable within the **Settings** menu, depending on the model chosen. As discussed before, if you are just starting out, then it is advisable to use the **Original** or **Lightweight** model.

  If you are starting a new model, then the model selected here will be the model used for all future training sessions of it. It is not possible to change a model type once it has been created.

The other options within this section can be ignored for your first model, although they may become more relevant as you gain experience using the software. As with all the options, the tooltips will tell you what these additional options do.

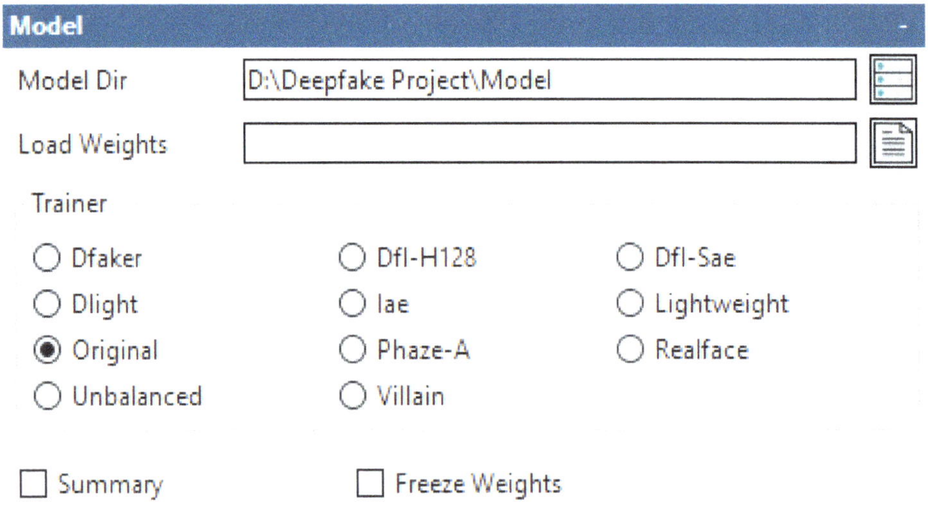

Figure 4.15 – The Model options for training in Faceswap

Training

These options relate to how the model should be trained:

- **Batch Size**: The number of faces to be fed through the model at once. Generally, a higher batch size will lead to a higher training speed; however, higher batch sizes will mean the model generalizes more. Increasing batch size is only sensible up to a limit. Anything beyond 128 and the model will start to struggle to obtain useful information for each batch.

  Batch size is also VRAM-limited. For more complicated models, you will have little choice but to use a smaller batch size, and obviously, the less VRAM available on the GPU, the more limited you are.

- **Iterations**: This can be left at default unless it is desired that the model should stop after a certain number of iterations. Knowing when to stop a model comes from experience and is dictated by the quality of output, so it will never be after a "set number of iterations."

- **Distributed**: This option is for multi-GPU users only. It allows for multiple video cards to be used, speeding up training by splitting batches over multiple devices.

> **Tip**
> It can be beneficial to start training a model at a higher batch size to get the speed benefits, and then reduce it later in training to get the benefits of drilling down for details.

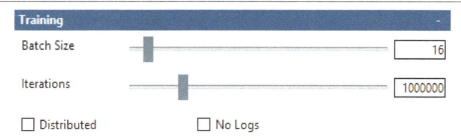

Figure 4.16 – The Training options within Faceswap

Augmentation

The **Augmentation** section allows you to enable or disable certain image augmentations (the way a neural network artificially increases the number of training images). When starting a new model, these should all be disabled (all augmentations are active). Later in the training session (when faces are becoming identifiable and more detailed), it can be desirable to turn some of these augmentations off:

- **Warp to Landmarks:** This is just an alternative warping technique. There is no conclusive evidence that enabling or disabling this option makes any real difference, so it is recommended to leave it disabled at all times.

- **No Flip**: Faces are vaguely symmetrical. The neural network leverages this knowledge by flipping around 50% of the images horizontally. For nearly all use cases, this is fine, and all images can be flipped at all times. However, in cases when there are distinct details on one side of a face (for example, a beauty mark), then this option should be enabled when you see that faces start to take shape within the training preview window.

- **No Augment Color**: The color augmentation helps the neural network to match color and lighting between the A and B sides by artificially coloring and changing other visual attributes of the images it sees. Generally, this is always desirable and should be left on, but for some use cases, it can be desirable to disable this augmentation.

- **No Warp**: This is possibly the most important option within this section. Warping is incredibly important to how a neural network learns. It is absolutely imperative that all models commence with warping enabled (failure to do so will invariably lead to a sub-standard model or, worse, model collapse). However, later in training, particularly when attempting to drill down into finer details, this warp actually becomes detrimental to the model's training, and so this option to disable the warp should be selected.

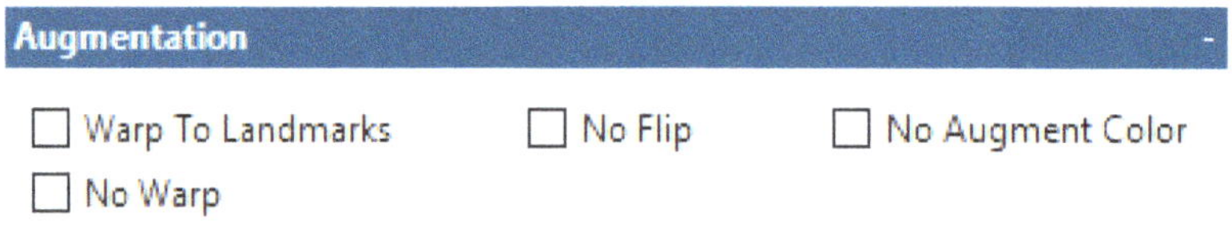

Figure 4.17 – The Augmentation section to train a model

- A rule of thumb is that if faces are recognizable and they do not appear to be getting any sharper over a significant period of time, then it is probably time to disable warping. Seeing clearly defined teeth and eye glare is a good indicator. It is important to note that it is near impossible to disable warping too late, but it is very definitely possible to disable warping too early, so err on the side of caution.

Author's note

Color augmentation and warping images are both invaluable ways to get more mileage from your data. Like the other augmentations in this section, they change your image slightly, as a way to effectively get new images that the AI model hasn't seen before.

Color augmentation works by slightly altering the colors of the image. This gives the model new colors to work with. This also helps the model with new lighting situations that might be absent from the data.

Warping works by slightly modifying the shape of the face in the image. This helps if certain expressions are less common in your data. It also helps ensure that the decoder builds the face from memory and not just from a copy of the original image.

Launching and monitoring training

Once the model configuration has been entered and all the settings have been adjusted to their appropriate value, the **Train** button can be pressed to launch the training session. Training can and will take a long time to complete, and there is no mathematical or numerical measure to know when a model has finished training. Knowing when a model is unlikely to improve anymore mostly comes with experience; however, there are some indicators in place that can help us to determine whether it is time to stop training.

Previews

Probably the most important measure of the model's progress is the preview images themselves. At each saved iteration, a series of preview images is generated to enable visualization of how the model is progressing.

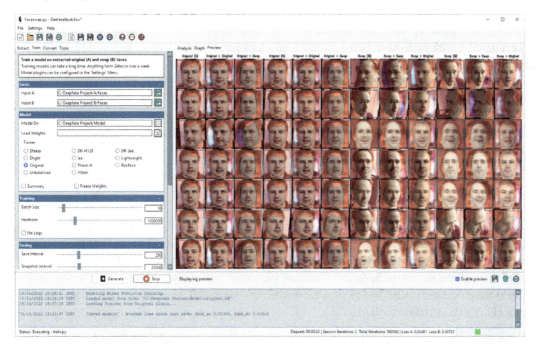

Figure 4.18 – Faceswap's GUI with the Preview window on the right

This preview consists of 28 faces randomly picked from the training set (14 for each side). For each training image, the original is shown, followed by an image showing the AI's attempt to recreate the original face. The final image shows the AI's attempt to swap the original face with the identity from the other side.

When training commences, these images may look like a solid color or a blurry image that is vaguely face-shaped; however, it will fairly quickly resolve into an identifiable face and slowly improve over time. It is worth noting that model improvement is not linear. While the faces will improve fairly quickly at first, this improvement will slow down until no visible difference will be seen from iteration to iteration. However, over a period of time, the model will improve. When comparing previews, it is not uncommon to compare the improvements of images that have been taken 10,000 to 100,000 iterations apart.

Loss

Another measure available to us is the loss for each iteration. Every time a batch is processed through the model, the neural network scores itself for how well it thinks it has recreated the images. While the loss value can swing wildly from batch to batch, over time the average value will drop. It is important to note that the actual value of the loss is not important. In fact, the value itself is effectively meaningless. The only issue to concern ourselves with is whether the value is dropping. This is for a couple of reasons; firstly, different loss functions will result in different numbers being generated, which are not comparable with each other. Secondly, the loss values given do not actually represent a score for anything that is useful for us to measure. The loss is generated by how well the neural network thinks it is recreating the **A** and **B** faces. It does this by looking at the original face and its recreation, and scoring itself based on the quality of the recreation. However, this score is not useful to us. What we would like to see is a score based on how well the neural network is taking a face and swapping it with another face. As there are no real-world examples of people who have had their faces swapped, this is an impossible measure to achieve, so we make do with using the loss values for face reconstruction rather than for face swapping.

Graphs

While loss on its own and at any given time is not a useful measure, its trend over time is. Ultimately, if loss is decreasing, the model is learning. The further a model is trained, the harder it is to ascertain whether the loss is actually still decreasing over time. Faceswap collects logging data for each batch passed through the model in the form of **TensorFlow event logs**. These logs are stored in the same folder as the model and can be used to visualize data in Faceswap's GUI, or analyzed using **TensorBoard** (TensorFlow's visualization toolkit).

To analyze the learning progress of an existing model, navigate to the **Analysis** tab on the right-hand side of the GUI. If session data is not already loaded, load a model's data by selecting the **Load** icon below the **Analysis** window and browse for the model's state.json file, located within the model folder. Once the data is parsed, session summary statistics will be displayed for each training session carried out:

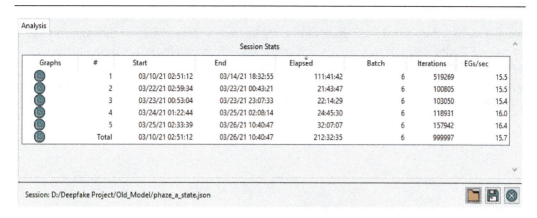

Figure 4.19 – Statistics for a series of Faceswap training sessions

Clicking the green **Graph** icon next to any of the session rows will bring up the training graph for that session, displaying the loss for each iteration during it:

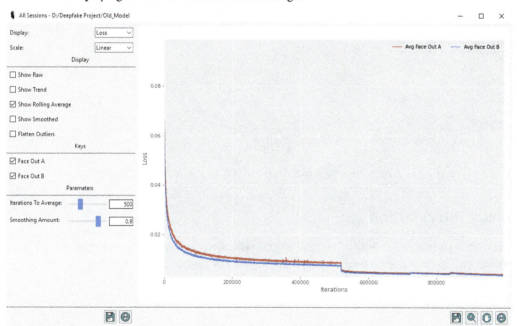

Figure 4.20 – A training graph showing loss over time for a Faceswap training session

For long sessions, it can be hard to ascertain whether the loss is still falling, due to the sheer quantity and range of data. Fortunately, it is possible to zoom into a selected range of the graph to get a better idea, by selecting the **Zoom to Rectangle** button toward the bottom right of the screen and selecting the area of interest. In this example, we shall zoom in on the last 100,000 iterations trained and make sure that we are viewing the rolling average of the loss values. As we can see, the loss is still clearly improving.

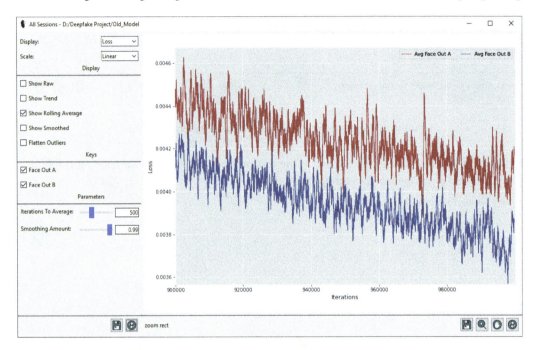

Figure 4.21 – A zoomed-in view of the last 100,000 iterations of a training session

Manual intervention

While training a model is mostly a "fire and forget" task, toward the end of training, it can help to take some steps to improve the final output. These steps are by no means necessary, but they can help with the final product. For any of these actions to take effect, the model will need to be stopped, the relevant settings adjusted, and then the model can be resumed:

- **Disabling Warp**: As previously mentioned, warping is imperative for a model to learn; however, as the model enters the middle to later stages of training, the warping augmentation can actually hurt the model in terms of image fidelity and clarity. Enabling the **No Warp** option is good practice for a high-quality final swap. As a general rule of thumb, this option should not be selected until you are at least 50% through the total train or you can see clearly defined features, such as individual teeth and eye glare. It is very hard to disable warping too late, but it is very easy to disable it too early. If the previews still look like they appear to be improving, then it is probably too early to disable warping.

- **Lowering Learning Rate**: As the model enters the late stage of training, it can help to lower the learning rate. This can help the model to drill down into the finer details. It is common to lower the learning rate a little, resume training, lower it some more, resume training, and repeat this cycle until you are happy with the results.

- **Fit Training**: A technique that can help is fitting data to the actual scene that is to be swapped. While it is not recommended to fully train a model only using data that will appear within the final swap, using this data can be useful to fine-tune an otherwise fully trained model.

When you are happy that you have trained the model as far as you can, it is time to take the trained model and apply it to a source video file to swap the faces.

Applying a trained model to perform a swap

Once the model has completed training, it can be used to swap the faces on any video to that contains the individual that is to be swapped out. Three items are required to successfully perform a swap – a video/series of images, a trained model, and an alignments file for the media that is to be converted. The first two items are self-explanatory; the alignments file is the one item we need to create.

The alignments file

The alignments file is a file bespoke to Faceswap, with a .fsa extension. This file should exist for every media source that is to be converted. It contains information about the location of faces within a video file, the alignment information (how the faces are orientated within each frame), as well as any associated masks for each frame.

Generating an alignments file is fairly trivial. In fact, at least one has been generated already when we built a training set. The process for generating training data and generating an alignments file is the same, bar a few changes. For this reason, a lot of the steps will be familiar to you, as they are the same ones we performed within the *Extracting faces from your source data* section. The most notable difference in this section is that alignment information needs to be generated for every single frame within the media source, while for generating a training set, it is more common to only extract faces for a subset of frames within the source material.

Please refer back to *Extracting faces from your source data* for a more detailed explanation of the common options between running an extract to generate training data and running an extract to perform a swap.

Within the Faceswap application, select the **Extract** tab, and then follow the following sections.

Data

This section is the location of the source material that we need to generate alignments for, as well as an output folder that faces will be extracted to. The faces generated here are not used by the Faceswap process at all, but they are very useful for cleaning our alignments file:

- **Input Dir**: The location of the media that we intend to swap the faces within
- **Output Dir**: The location that identified faces will be extracted to

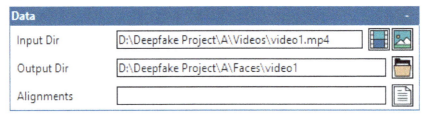

Figure 4.22 – The Data options within Faceswap's extraction settings

Plugins

The plugins that are selected are likely to be the same as those selected when generating a training set, so refer to the previous section for more information on the options. Only one option within this section will likely change when extracting an alignments file:

- **Re Feed**: As the extraction process has no understanding of temporal coherence (that is, how each frame relates to the previous and subsequent frame), it can lead to "jittery" alignments within the final swap. This means that the face in the final swap moves a small amount from frame to frame. While this is not important for training sets, it is important when generating the file for the final swap.

  Re Feed is a mechanism to help prevent this jitter by feeding a detected face into the aligner a set number of times and taking the average of the results. It is worth noting that each increase in the re-feed amount will slow extraction down, as the data needs to be passed through the process multiple times. The higher this number is set, the smoother the final output should be, but it is diminishing returns. Setting the value too high is unlikely to net any visible benefit but will take significantly longer to run an extract.

  Set this to a value that brings you satisfactory results, while also running at a speed you can live with. Any value above 1 should give improved results over extracting without re-feed. A re-feed value of 8–10 will likely get the output close to as good as it can be.

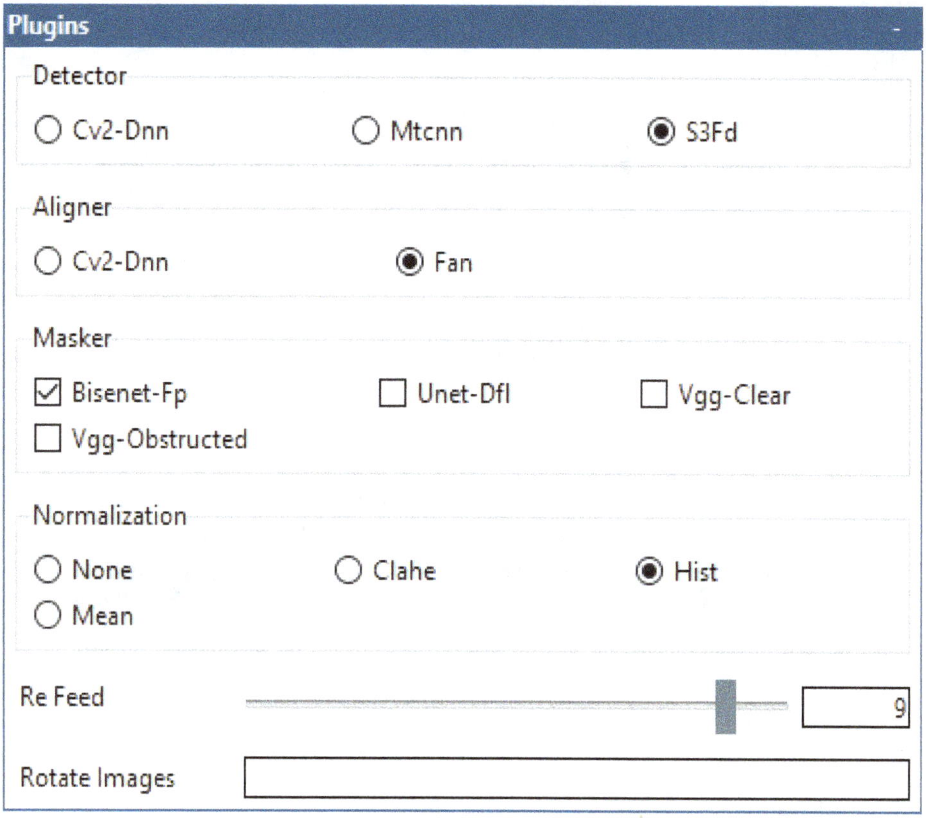

Figure 4.23 – The Plugins options within Faceswap's extraction settings

Face Processing

As with extracting training faces, **Min Size** is the only option within this section that may need to be adjusted. The value specified here will generally correspond to the size of the faces within the source material, but leaving it set at a low value, regardless, can help with removing some false positives that are clearly not valid faces.

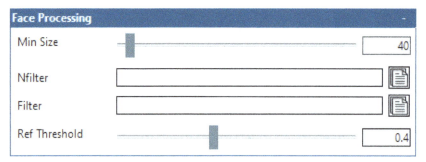

Figure 4.24 – The Face Processing options within Faceswap's extraction settings

Output

Finally, the **Output** section should be reviewed and updated. The only option that needs to be amended from the settings used for extracting a training set is **Extract Every N**. It is imperative that every frame has a corresponding entry in the generated alignments file, so this should be set to **1**.

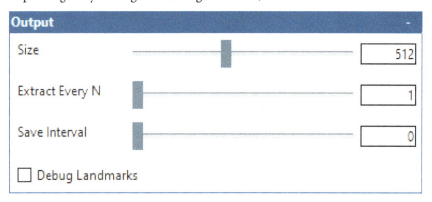

Figure 4.25 – The Output options within Faceswap's extraction settings

Extract

Once the appropriate settings have been locked in, press the **Extract** button, to generate the alignments file and extract the found faces into the given folder. The amount of time this will take will depend on the available hardware, the length of the source material, and the re-feed value that has been set.

Cleaning the alignments file

Similarly to when we collected faces to build a training set, the extraction process will have done a decent job of identifying faces, but it will not have done a perfect job, so some manual processing is now required to clean up the alignments file. The process is the same as that in the *Curating training images* section with an additional step, so follow the steps within that section to perform the initial cleansing of the alignments file. Once unwanted faces have been removed from the alignments file, the folder of extracted faces can be deleted. The faces are not actually used by the conversion process; they are just used as a mechanism to clean the alignments file.

At this point, we should have a source video that is the target for swapping faces and a corresponding alignments file that holds information about the location of faces within that file. It is entirely feasible to run a conversion at this stage, but another step is required for the best results – fixing the alignments file.

Fixing the alignments file

Whilst we have removed unwanted faces, false positives, and any clearly misaligned faces from our alignments file, some further work is required to clean up the file for the final conversion. The main reasons for this are to fix frames where the following scenarios occur:

- Multiple faces have been identified. Sometimes, the detector will find two faces in a frame, but the aligner performs alignment on the same face twice. This often happens when two faces appear close to each other within a frame.

- The face is not aligned correctly. Sometimes, the face may appear aligned correctly when scanning through the folder of images, but examining the landmarks will demonstrate that this is not the case. These faces will sometimes convert correctly, but often this misalignment will lead to a messy swap for those frames (the swap will not look quite correct, it may look blurry, or may flicker between frames).

- A face hasn't been identified. Ensuring that all faces being swapped have been identified is necessary; otherwise, the original face, rather than the swapped face, will appear in those frames.

- The mask has not been detected correctly. Depending on the conditions of the source frame, some neural network-based masks may not have been detected correctly, so these need to be fixed up. Depending on how the mask has been rendered, an incorrect mask may mean that parts of the original face show through, or parts of the background frame do not render correctly.

Again, Faceswap provides tools to make this process easier – specifically, the **Manual tool**, which enables the visualization and editing of alignments/masks within the context of the original frame.

To launch the Manual tool, navigate to the **Tools** tab and then the **Manual** sub-tab:

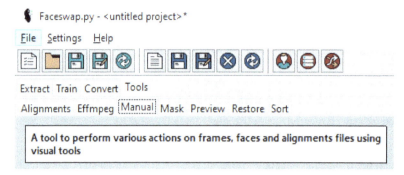

Figure 4.26 – The location of the Manual tool within Faceswap's GUI

Assuming that the alignments file for the video to be converted is in the default location, then only one argument needs to be provided to launch the Manual tool the location of the source video/folder of images that is to be converted. Specify this location within the **Frames** box and hit the **Manual** button to launch the tool.

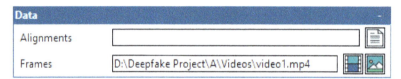

Figure 4.27 – The Data options for launching Faceswap's Manual Tool

Once the tool loads, you will be greeted with a main "video" window that shows the source that is being worked on and a secondary window that displays all of the faces that exist within the alignments file.

Figure 4.28 – The Faceswap manual tool showing the video window
at the top and the faces viewer at the bottom

The top left buttons allow you to perform different actions on the alignments, so hover over the tooltips to see what is available:

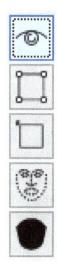

Figure 4.29 – The Editor selection buttons within Faceswap's Manual tool

Another area of interest is the **Filter**. The Filter is a pull-down list that is located between the frames and faces windows and enables you to filter the frames and faces shown in the tool by certain criteria:

Figure 4.30 – The face Filter options within Faceswap's Manual tool

Finally, the faces window also has some buttons on the left-hand side, which enable you to toggle the display of the face landmark mesh and the selected mask for faces displayed in the faces area:

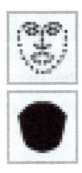

Figure 4.31 – The face viewer buttons within Faceswap's Manual tool

Removing multiple faces

To remove any extra faces from frames that contain multiple faces, select the **Multiple Faces** filter. The main video window will now only show any frames that contain multiple faces. If no frames show in the top window and no faces show in the bottom window, then there are no frames with multiple faces, and you can move on to the next action.

Knowing which face to remove is sometimes not obvious, so press the **Display landmarks mesh** button to the left of the faces window to bring up the landmarks overlay. If none of the faces are obviously misaligned (when the displayed landmarks mesh does not correspond to the underlying facial features), then any of the multiple faces can be deleted; otherwise, aim to delete the face with the landmarks that correspond least with the underlying face.

There are several ways to delete faces from the frame. They can be deleted from the main video window by hovering over the unwanted face and pressing the **Del** key, or by right-clicking on the unwanted face in either the video or faces window and selecting **Delete Face**. When a frame no longer contains multiple faces (just one face remains in the frame), the faces for that frame will be removed from the faces window. Using this mechanism, it is usually quickest to right-click and select **Delete Face** from the faces window for all those faces that are unwanted until no faces remain.

Once all frames with multiple faces have been cleaned, hit the **Save** icon to save the changes to the alignments file.

Fixing misaligned faces

The Manual tool has misalignment detection built in. It is not perfect, but it does help in identifying and fixing the most obviously misaligned faces. While the detection can find faces that are obviously misaligned, it will not find faces where the face landmarks are in the correct location in relation to each other but do not correspond with the underlying face.

Select the **Misaligned Faces** filter to only display frames and faces where misaligned faces have been detected. A slider will appear next to the filter list box to control the distance:

Figure 4.32 – The Distance slider for selecting misaligned faces

This is how far the landmark points within each face have to be from an "average face" to be considered misaligned. Low values will be less restrictive, so are likely to contain faces that are properly aligned but are at more extreme angles/poses. Generally, distances between 6 and **10** work fairly well. A distance of **10** should only show misaligned faces. A distance of 6 is likely to show a mixture of misaligned faces and more extreme poses, but it will catch misaligned faces that higher values will miss. It can be easier to set a higher distance (**10**, for example) and fix the misaligned faces that appear. Then, set the distance to 8 and repeat the process, continuing to step down until the filter is not catching enough misaligned faces in relation to more extremely posed faces.

Regardless of the distance that has been set, to fix misaligned faces, the following actions should be taken:

1.   Enable the landmark mesh for the faces viewer by toggling the landmarks button to the left of the faces viewer.

2.   Click on a face that has misaligned landmarks.

3.   Within the main frame editor, make sure that the Bounding Box Editor is selected. This editor allows for control of the blue box around the face. This is the "detected face" box that was picked up by the face detector during the extraction phase. Adjusting this box will update the face that is fed to the aligner, with new landmarks being calculated. Continue to adjust the box until the landmarks align correctly.

4.   If the landmarks are not aligning, it can help to switch between the different **Normalization Methods** options in the right-hand settings box. Different methods work better or worse in different situations, but **Hist** or **Clahe** tend to return the best results.

5. Some faces can be stubborn (difficult angles, obstructions, or bad lighting). In these cases, it can be next to impossible for the aligner to detect the landmarks. A couple of other editors can be used in these situations:

    - **Extract Box Editor**: This editor shows a green box that corresponds to the area of the frame that will be extracted if face extraction is run. It is possible to move, resize, and rotate this extract box, which will impact the location of the landmarks within the extract box. This can be leveraged to quickly align a face by copying the landmarks from the previous or next frame (assuming that the landmarks have not changed too much between frames – for example, a scene change) and quickly adjusting the extract box to fit the current frame.

    - **Landmark Point Editor**: This editor allows for the location of each individual point within the 68 landmarks to be manipulated. This level of granular control is rarely necessary, but it exists if it is needed.

Once the obviously misaligned landmarks have been fixed, hit the **Save** button to update the changes to the alignments file.

Adding missing faces

Some frames may not have had faces identified where they should have been. The most common reason for this is that the detector did find a face, but the aligner failed to align it correctly, and then the face was deleted during the sorting process. Again, the Manual tool has a filter to help with this.

Select the **No Faces** filter to filter the top window to only those frames where no faces appear. The bottom window will remain empty for this particular filter. Navigate through the video until a frame that contains a face that has not been detected is reached, and make sure that the Bounding Box Editor is selected.

Landmarks can be created by clicking over a face within a frame, or copied from the previous or next frame and then amended. The bounding box can then be edited in the same way as in the previous step, with the same caveat about difficult faces.

When all frames that were missing faces have been fixed, hit the **Save** button to save the changes to the alignments file.

Final alignments fixups

Once the obvious missing and misaligned faces have been fixed, it's time to perform any final fixes to the alignments file. This is as simple as scrolling through all of the faces in the faces viewer and fixing any faces that remain misaligned. The faces viewer window can be expanded to show more faces within a single screen, and then the page-up/page-down buttons can be used to scroll through the faces a page at a time. When a misaligned face is discovered, it can be clicked on, and then the Bounding Box Editor can be used to re-align the face correctly.

Finally, once all faces have been reviewed and fixed, press the **Save** button to save the final changes to the alignments file.

Now that the alignments file has been fixed, you can close the Manual tool.

Regenerating masks

If an neural network-based mask is to be used for the swap, then these masks will need to be re-generated for any faces where the alignment data has been edited. The reason for this is that the aligned face, generated from the landmarks, is used to generate the face that is fed into the masking model. Once these landmarks have been edited, the mask is invalidated, so the process automatically deletes these invalid masks when the alignments are changed.

Again, Faceswap provides a tool to add masks to existing alignments files – the appropriately named **Mask tool**. This tool can be used to generate masks that did not previously exist in the alignments file, regenerate all masks, or just populate masks for those faces that are missing the specified mask:

1. Navigate to the **Tools** tab and then the **Mask** sub-tab:

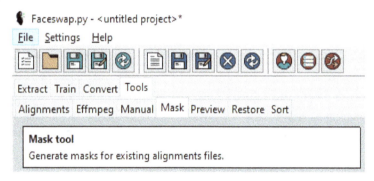

Figure 4.33 – The location of the Mask tool within Faceswap's GUI

2. Within the **Data** section, add the path to the video file to regenerate masks for the **Input** field, as well as the corresponding alignments file for the **Alignments** field. As the source to be worked on are the final frames to swap onto, make sure that **Frames** is selected under **Input Type**:

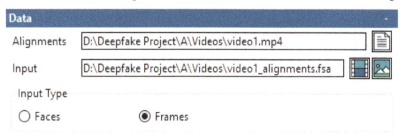

Figure 4.34 – The Data settings of Faceswap's Mask tool

3. In the **Process** section, select the mask that is to be populated into the alignments file for **Masker**. Under **Processing**, select **Missing** if masks have already been generated and the goal is to repopulate those masks that are associated with faces that have had their alignments fixed; otherwise, select **All** to generate masks for every face within the alignments file:

Figure 4.35 – The Process settings of Faceswap's Mask tool

4. The **Output** section is just for visualizing the masks, serving no practical purpose, so it can be ignored.

5. Press the **Mask** button to generate the missing masks and save them to the alignments file.

Fixing masks

A final optional step is to fix up the generated masks. Much like face alignment, the neural networks that generate the masks are good, but they are often not perfect. This can be down to a number of reasons, such as lighting conditions and the quality of an image. In particular, obstructions in front of the face are not handled well by any of the maskers, so these will need to be manually edited.

This should be the absolute last action performed on the alignments file. Any edits performed on landmark data within the alignment file will strip any neural network masks from the file and overwrite any edited landmark-based masks with the latest landmark data, destroying any manual edits that have been performed.

The Manual tool, used to fix up the alignments, can also be used to fix masks:

1. Launch the Manual tool by selecting the **Tools** tab, followed by the **Manual** sub-tab, and launch in the same way as before.

2. Select the **Mask Editor** button from the buttons next to the frame viewer, and then select the mask type to be edited from the right-hand side options panel.

3. Press the **Mask Display** toggle button next to the faces viewer to display the selected mask within the faces window.

4. Scroll through the faces window, looking for masks that require fixing. If a face is discovered that requires editing, it can be clicked on to bring the relevant frame into the frame viewer. The **Brush** and **Eraser** tools can then be used to paint in or out the desired mask areas.

Once all the masks have been fixed, press the **Save** button to save the mask edits to the alignments file.

Using the Preview tool

It is possible to process the swap now and view the final output. However, some settings will need to be adjusted on a case-by-case basis, specifically various post-processing actions, such as mask erosion/blending, color correction, and sharpening.

Faceswap includes the **Preview tool** to help lock these settings in prior to running the final conversion, which can be accessed by selecting the **Tools** tab and then the **Preview** sub-tab:

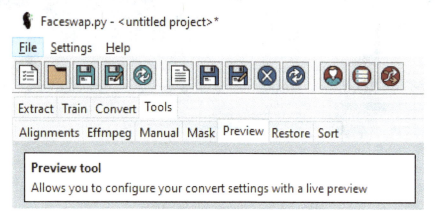

Figure 4.36 – The location of the Preview tool within Faceswap's GUI

To launch the tool, provide the location of the video you intend to swap onto in the **Input Dir** field, and the folder that contains the trained model in the **Model Dir** field. The **Alignments** field can be left blank, unless the alignments file has been moved or renamed, in which case it will need to be explicitly specified:

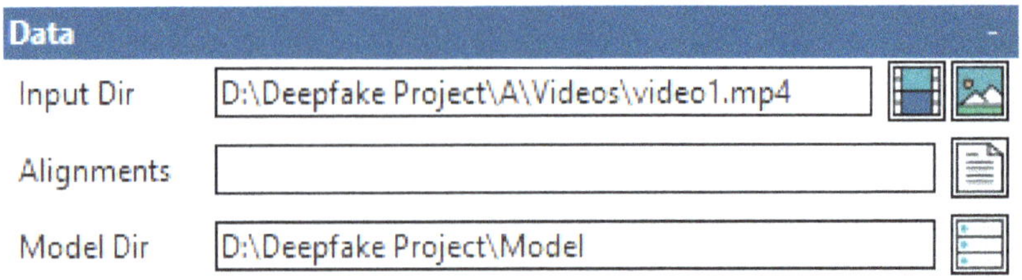

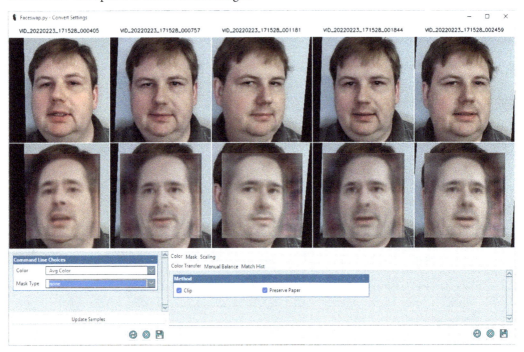

Figure 4.37 – The Data settings for Faceswap's Preview tool

Press the **Preview** button to launch the tool.

The tool is split into three sections. The main window shows the faces from the original frame in the top row, with the swap applied with current settings in the bottom row. As settings are adjusted, the bottom row will update to reflect these changes.

Figure 4.38 – Faceswap's Preview tool

The bottom-left panel displays command-line choices, while the bottom-right panel displays plugin settings.

Command-line choices

These are parameters that are chosen each time the conversion process is run (these options are not persistent), so you will need to remember what is set here to replicate it in the main Faceswap conversion process. Specifically, the options that can be set here are as follows:

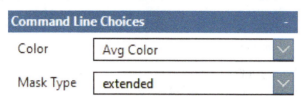

Figure 4.39 – The Preview Tool's Command Line Choices

- **Color**: The color-matching methodology to use. The best choice here will depend on the scene being converted. The **Color Balance**, **Manual Balance**, and **Match Hist** options have further configuration options that can be adjusted from the **Plugin Settings** section.

- **Mask Type**: The type of mask to use for overlaying the swapped face onto the original frame. By default, the landmarks-based **extended** and **components** masks will be available. Any additional neural network-based masks that exist within the alignments file will also be accessible, as well as the option to entirely disable the mask (**None**). The settings that control the blending of the chosen mask into the background frame are adjusted from the **Plugin Settings** section.

Plugin settings

This section contains the configuration settings for the various post-processing plugins available in Faceswap. Values selected here, once saved, are persisted for all future conversions. As such, unlike the **Command Line Choices** options, there is no need to make a note of what is being set within this section.

The plugin settings are split into three configuration groups:

- **Color**: Configuration options for color-matching plugins. The actual methodology to use is selected within the **Command Line Choices** section, but some of the choices have additional configuration parameters that are controllable here. Make sure that the correct methodology is set within the **Command Line Choices** section to observe any changes within the main window.

- **Mask**: Options to control the blending of the swapped image into the background frame. These settings are broken down into two further categories – **Box Blend**, which controls the settings that blend the extracted square containing the face into the background frame, and **Mask Blend**, which controls the settings for blending the mask around the face into the background frame. (Note that if **None** has been selected as the mask type in the **Command Line Choices** section, then any changes made within the **Mask Blend** settings will not be visible within the preview window.)

How much of an impact each of these settings will have on the final output will depend greatly on the coverage and centering options that were selected when training the model. For example, with a low coverage and legacy centering (that is, very closely cropped), it is entirely possible that an extracted face box is contained entirely within the mask, in which case **Mask Blend** settings would have no visible effect. Similarly, with high coverage, and face or head centering, it is possible that the full mask exists within the extract box, in which case the **Box Blend** settings would have no visible effect. In most cases, adjusting a combination of the two blending settings will be necessary.

- **Scaling**: The final configurable plugin controls any artificial sharpening to apply to an image. Quite often, the swapped face will need to be upscaled to fit into the final frame. This section allows you to control any sharpening effects to help better upscale the image.

> Tip
>
> To get a better impression of the effects of adjusting the mask plugin settings, select **Manual Balance** as the color command-line choice, and then adjust **Contrast** and **Brightness** to **–100** within the **Manual Balance** plugin setting. This will display the swap area as entirely black, which can make it easier to adjust the mask correctly.

Tuning the conversion settings

The actual configuration choices to be used will vary on a video-by-video basis; there are no hard and fast rules, so it is just a question of adjusting the settings until a satisfactory result is achieved. Once a plugin is configured correctly, that plugin's configuration can be saved by clicking the bottom-right **Save** button. To save the settings for all plugins that have been adjusted, click the bottom-left **Save** button.

When appropriate settings have been locked in, make a note of the **Command Line Choices** settings and exit the Preview tool.

Generating the swap

Once the model has been trained, the alignments file has been created, and the swap settings have been locked in, the final product can be created. The process of generating a swap is called **converting** – that is, converting the faces in a source video from their original form to the version generated from the trained model.

Converting is probably the least involved of the main Faceswap processes. Access the **Convert** section of the Faceswap application by selecting the **Convert** tab:

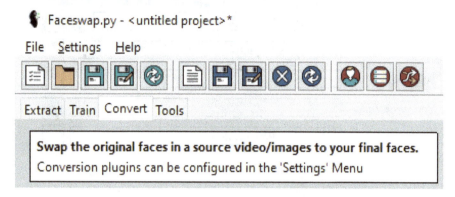

Figure 4.40 – The location of the Convert settings within Faceswap's GUI

Data

This section is used to tell the process where the assets are located to perform the swap, as well as where the final output should be exported to:

- **Input Dir**: The location of the source video or folder of images to be processed.

- **Output Dir**: The location that the converted media should be outputted to. This folder should not pre-exist.

- **Alignments**: Optionally, specify the location of the alignments file. If the alignments file is in its default location (next to the source video) with the default name, then this can be left blank, as the file will be detected.

- **Reference Video**: This option is only required if the source is a folder of individual frames and the desired output is a video file. The reference video would be the original video file that the folder of frames was extracted from, and it provides the conversion process with the audio track and the FPS that should be compiled into the final video.

- **Model Dir**: The location of the folder that contains the trained Faceswap model.

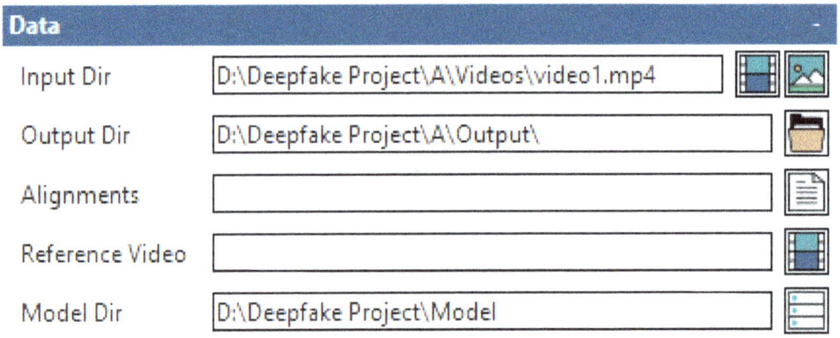

Figure 4.41 – The Data options within Faceswap's Convert settings

Plugins

There are several plugins available for the conversion process, which are selectable here. Two of the plugins will have been seen before when we used the Preview tool (**Color Adjustment** and **Mask Type**), so ensure that you select the same options here as those selected within the Preview tool for the output to remain consistent:

- **Color Adjustment**: The color correction to use. This will have been previewed and selected using the Preview tool, so select the same plugin here. The actual plugin to use will vary from project to project.

- **Mask Type**: The type of mask to use to overlay the swapped face over the original frame. The chosen mask here must exist within the alignments file (**Components** and **Extended** will always exist; other masks need to be generated). Generally, this will be the same mask that the model was trained with and will have been previewed with the Preview tool, so select the same mask that was used to preview.

- **Writer**: The plugin to use to create the final media. The writer plugins are used to generate the final product. **Ffmpeg** is used to create video files, **Gif** is used to create animated GIFs, and **OpenCV** and **Pillow** will create a folder of images, with OpenCV being quicker but having a more limited file format choice than Pillow.

The writers can each be configured by selecting **Settings | Configure Settings** and selecting the relevant writer plugin under the **Convert** node:

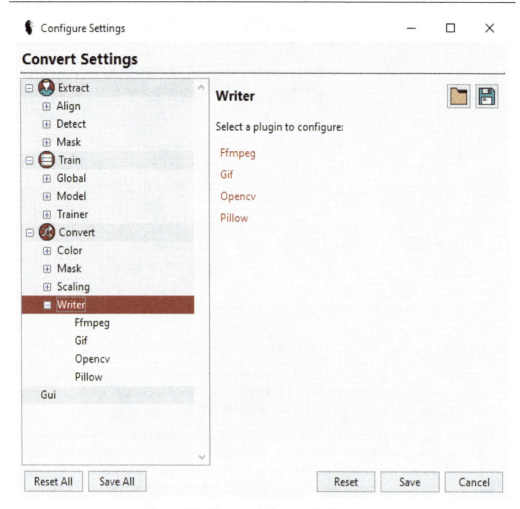

Figure 4.42 – Faceswap's Convert plugin settings

It is possible to create the swap on a separate transparent layer only containing the swapped face and the mask, to overlay over the original frame in various external VFX applications. The **OpenCV** and **Pillow** writers both support this, with OpenCV allowing the generation of four-channel PNG images and Pillow allowing the generation of four-channel PNG or TIFF images. This option can be enabled by selecting the **Draw Transparent** option within either of these plugins' configuration settings.

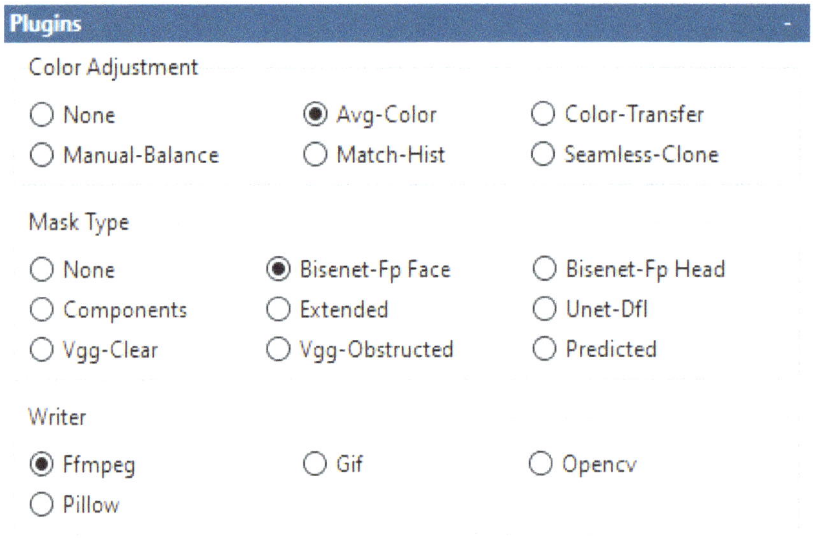

Plugins

Color Adjustment

○ None   ● Avg-Color   ○ Color-Transfer
○ Manual-Balance ○ Match-Hist   ○ Seamless-Clone

Mask Type

○ None   ● Bisenet-Fp Face   ○ Bisenet-Fp Head
○ Components   ○ Extended   ○ Unet-Dfl
○ Vgg-Clear   ○ Vgg-Obstructed   ○ Predicted

Writer

● Ffmpeg   ○ Gif   ○ Opencv
○ Pillow

Figure 4.43 – The Plugins options within Faceswap's Convert settings

Other settings

Frame Processing can be ignored in most cases. The most likely option of interest is **Output Scale** though, which scales the final output media by the designated amount. For example, setting an output scale of **50** for a 720p input video will result in a final output at 360p.

The **Face Processing** section can be ignored. If the alignments file has been created correctly, then none of the options here are relevant. Similarly, most of the options within the **Settings** section can be ignored in most cases. The only possible exception to this is the **Swap Model** option. This can be used to create a swap in the opposite direction to which the model was trained – that is, instead of **A** > **B**, the conversion will run **B** > **A**. This can be useful if you have a model trained on a face pair and you wish to run conversion in the opposite direction, or if a model has accidentally been trained the wrong way around (the original face has been trained on the **B** side, with the desired swap trained on the **A** side).

Once all of the settings are set correctly, press the **Convert** button to apply the trained model on your source media and generate the swap in the output destination.

Summary

In this chapter, we learned the workflow required to create a deepfake using the open source Faceswap software. The importance of data variety was discussed and the steps required to acquire, curate and generate face sets were demonstrated. We learned how to train a model within Faceswap, and how to gauge when a model has been fully trained, as well as learned some tricks to improve the quality of the model. Finally, we learned how to take our trained model and apply it to a source video to swap the faces within the video.

In the next chapter, we will begin to take a hands-on look at the neural networks available to build a deepfake pipeline from scratch using the PyTorch ML toolkit, starting with the models available for detecting and extracting faces from source images.

Part 2: Getting Hands-On with the Deepfake Process

This part of the book is all about getting hands-on with the code. We will look deep into exactly what it takes to make a deepfake from beginning to end, leaving no stone unturned or line of code unexplained. If you're here for the code, this is the section for you.

In the first chapter of this section, we'll examine extraction. This is the process of getting all the faces out of a video so that we can use them in other stages of the process. We'll look at the process of turning a video into frame images, then we'll go through all the code necessary to turn the frames into clean, aligned faces with matching mask images ready for training. After that, we'll look into training, examine the neural network from the bottom up, and then show the entire learning process of the model. Finally, we'll get into conversion, where we'll examine the process of going through every image to swap a new face onto the original, including turning it back into a video.

By the end of this part, you'll know exactly how to code your own deepfakes from beginning to end.

This part comprises the following chapters:

- *Chapter 5, Extracting Faces*
- *Chapter 6, Training a Deepfake Model*
- *Chapter 7, Swapping the Face back into the Video*

5

Extracting Faces

In this chapter, we will start our hands-on activities with the code. To begin with, we'll cover the process of extracting faces.

Extracting faces is a series of steps in a process that involves many different stages, but it's the first discrete stage in creating a deepfake. This chapter will first talk about how to run the face extraction script, then we will go hands-on with the code, explaining what each part does.

In this chapter, we will cover the following key sections:

- Getting image files from a video
- Running extract on frame images
- Getting hands-on with the code

Technical requirements

To proceed, we recommend that you download the code from the GitHub repo at `https://github.com/PacktPublishing/Exploring-Deepfakes` and follow the instructions in the `readme.md` file to install Anaconda and to create a virtual environment with all required libraries to run the scripts. The repo will also contain any errata or updates that have happened since the book was published, so please check there for any updates.

Getting image files from a video

Videos are not designed for frame-by-frame access and can cause problems when processed out of order. Accessing a video file is a complicated process and not good for a beginner-level chapter like this. For this reason, the first task is to convert any videos that you want to extract from into individual frames. The best way to do this is to use **FFmpeg**. If you followed the installation instructions in the *Technical requirements* section, you will have FFmpeg installed and ready to use.

Let's begin the process.

> **Tip**
>
> When you see code or command examples such as those present here with text inside of curly brackets, you should replace that text with the information that is explained in the brackets. For example, if it says `cd {Folder with video}` and the video is in the `c:\Videos\` folder, then you should enter `cd c:\Videos`.

Place the video into a folder, then open an Anaconda prompt and enter the following commands:

```
cd {Folder with video}
mkdir frames
ffmpeg -i {Video Filename} frames\video_frame_%05d.png
```

This will fill the `frames` folder with numbered images containing the exported frames.

There are a lot of options you can control with FFmpeg, from the resolution of the output image to the number of frames to be skipped. These features are beyond the scope of this book but have been covered extensively elsewhere. We advise searching for a guide to ffmpeg's command line options if you're going to do much more than the basics.

Running extract on frame images

To run the extract process on a video, you can run the extract program from the cloned `git` repository folder. To do this, simply run the following in an Anaconda prompt:

```
cd {Folder of the downloaded git repo}\
python C5-face_detection.py {Folder of frame images}
```

This will run the face extraction process on each of the images in the folder and put the extracted images into the `face_images` folder, which (by default) will be inside the folder of frame images. This folder will contain three types of files for each face detected, along with a file containing all alignments.

face_alignments.json

There will just be one of these files. It is a JSON-formatted file containing the landmark positions and warp matrix for every face found in the images. This file is human-readable like any JSON file and can be read or edited (though it's probably not something that you'd do manually).

face_landmarks_{filename}_{face number}.png

This is a copy of the original image with a bounding box drawn around the face and five **landmark** points written out. Landmarks are common points on the face. We'll cover the usage of these landmark points later, but the ones we're most interested in are the eyes, nose, and corners of the mouth.

Figure 5.1 – Example of face_landmarks_bryan_0.png

Author's note

All of the images of people used for practical demonstration in this section are of the authors.

face_bbox_{filename}_{face number}.png

This image represents the original **bounding box** (the smallest box that surrounds the detected face) found in the original image. It will be in the full original size and angle of the face found in the image.

Figure 5.2 – Example of face_bbox_bryan_0.png

face_aligned_{filename}_{face number}.png

This image will be a smaller size (the default is 256x256) image of the face. This is an **aligned** face image, where the face has been lined up according to the landmarks. This image should generally have the face centered in the image and lined up vertically. If the face in the bounding box image was crooked or angled in the box, it should be straightened in the aligned image. There may also be a black cutout where the edge of the original frame cuts off the data.

This image is the most important image and is the one that will be used for training the model. It is critical for quality training that the aligned face is a good-quality image. This is the data that you'll want to clean to get a successful deepfake.

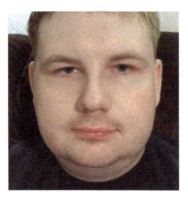

Figure 5.3 – Example of face_aligned_bryan_0.png

face_mask_{filename}_{face number}.png

This image matches the size of the aligned image. In fact, it matches the aligned image in the crop, rotation, and size. The mask is an AI-predicted outline of the face that will be used later to ensure that the face is trained properly and help swap the final face back onto the image.

Figure 5.4 – Example of face_mask_bryan_0.png

The mask should line up with the aligned face perfectly, showing where the face and edges are. You can see how the mask overlays onto the face in *Figure 5.5*.

Figure 5.5 – Example of mask overlayed on aligned face

So, now that you know how to run the face detector and all of its output, let's get into the hands-on section and determine exactly what it's doing.

Getting hands-on with the code

Now it's time to get into the code. We'll go over exactly what C5-face_detection.py does and why each option was chosen. There are five main parts to the code: initialization, image preparation, face detection, face landmarking/aligning, and masking.

Initialization

Let's begin.

> **Author's note**
>
> Formatting for easy reading in a book requires modifying the spacing in the samples. Python, however, is whitespace sensitive and uses spacing as a part of the language syntax. This means that copying code from this book will almost definitely contain the wrong spacing. For this reason, we highly recommend pulling the code from the Git repository for the book at https://github.com/PacktPublishing/Exploring-Deepfakes if you plan on running it.

1. First, we import all required libraries:

```
import os
import torch
import cv2
```

```
import json_tricks
import numpy as np
from tqdm import tqdm
from argparse import ArgumentParser
from face_alignment.detection.sfd import FaceDetector
from face_alignment import FaceAlignment, LandmarksType
from skimage.transform._geometric import _umeyama as
umeyama
from lib.bisenet import BiSeNet
```

In this section, we're importing all the important libraries and functions that we will use. These are all used in the code and will be explained when they're used, but it's a good idea to get familiar with the imports for any project since it lets you understand the functions being used and where they came from.

Many of these imports are from the Python standard library, and if you're interested in reading more about them, you can find their documentation online. We'll be explaining those that aren't part of the standard library as we come to them, including where you can find documentation on them.

2. Next, let's skip to the end of the code for a moment, where we'll look at argument parsing:

```
if __name__ == "__main__":
""" Process images in a directory into aligned face
images
Example CLI:
------------
python face_detection.py "C:/media_files/"
"""
parser = ArgumentParser()
parser.add_argument("path",
    help="folder of images to run detection on")
parser.add_argument("--cpu", action="store_true",
    help="Force CPU usage")
parser.add_argument("--size", default=256,
    help="height and width to save the face images")
parser.add_argument("--max_detection_size", default=1024,
    help="Maximum size of an image to run detection on.
      (If you get memory errors, reduce this size)")
parser.add_argument("--jpg", action="store_true",
```

```
        help="use JPG instead of PNG for image saving
            (not recommended due to artifacts in JPG images)")
    parser.add_argument("--min_size", default=.1,
        help="Minimum size of the face relative to image")
    parser.add_argument("--min_confidence", default=.9,
        help="Minimum confidence for the face detection")
    parser.add_argument("--export-path",
        default="$path/face_images",
        help="output folder (replaces $path with the input)"
    opt = parser.parse_args()
    opt.export_path = opt.export_path.replace('$path',opt.
    path)
    main(opt)
```

In this section, we define the options available in the script. It lets you change many options without needing to modify the code at all. Most of these will be fine with the defaults, but the argument parser included in the Python standard libraries gives us an easy-to-use method to make those changes at runtime. The documents and guides for the argument parser are part of the standard library, so we'll skip over the basics, except to state that we use `parser.add_argument` for each argument that we want, and all the arguments get put into the `opt` variable.

One special thing we're doing here is to change the `export_path` variable after it is defined, replacing the `$path` string in the variable to put the output folder into the input folder. If the user overrides the default, there will be nothing to replace (or the user could use `$path` to specify a different subfolder in the input folder if desired).

> **Note**
> Python norms require that this section is at the end of the file to work. However, it's very important to at least look at this section so that you know what it is doing before we go through the rest of the code.

3. Next, we return to the following code:

    ```
    def main(opt):
    ```

    This section begins with the main function we're using to do all the actual work.

4. The first thing we do is make sure the output folder exists:

    ```
    if not os.path.exists(opt.export_path):
        os.mkdir(opt.export_path)
    ```

    This creates the output folder if it doesn't already exist.

5. Then we enable **graphics processing unit (GPU)** acceleration if it's available:

    ```
    device = "cuda" if torch.cuda.is_available() and not opt.
    cpu else "cpu"
    ```

    In this part, we begin with the **artificial intelligence (AI)** work. In this case, we determine whether there is an NVIDIA GPU and if there is we use it for hardware acceleration unless the user has disabled it with a command line switch. The hardware acceleration layer we use is provided by NVIDIA and is called **CUDA**. We check with PyTorch for CUDA availability and save the device variable for later use.

6. Next, we'll initialize the face detector and aligner:

    ```
    face_detector = FaceDetector(device=device,
    verbose=False)
    face_aligner = FaceAlignment(LandmarksType._2D,
    device=device, verbose=False)
    ```

    This code defines the face detector and aligner that we'll be using; it specifies what devices we're going to use and tells the aligner to use a model trained for 2D face landmarks. These classes come from the face_alignment library available at https://github.com/1adrianb/face-alignment. Both the detector and the aligner are AI models that have been trained for their specific uses. The detector finds any faces in a given image, returning their position through the use of a bounding box. The aligner takes those detected faces and finds landmarks that we can use to align the face to a known position.

7. Next, we define our masker and prepare it. We load the pre-trained weights and set it to use CUDA if NVIDIA support is enabled:

    ```
    masker = BiSeNet(n_classes=19)
    if device == "cuda":
      masker.cuda()
    model_path = os.path.join(".", "binaries",
      "BiSeNet.pth")
    masker.load_state_dict(torch.load(model_path))
    masker.eval()
    desired_segments = [1, 2, 3, 4, 5, 6, 10, 12, 13]
    ```

    We use the desired_segments variable to define which of the masker segments we want to use. In our current masker, this gives us the face itself and discards background, hair, and clothing so that our model will only need to learn the information that we want to swap.

8.   Lastly, we get a list of files in the input folder:

```
alignment_data = {}
list_of_images_in_dir = [ file for file in
os.listdir(opt.path) if os.path.isfile(os.path.join(opt.
path,file)) ]
```

This code first prepares an empty dictionary to store the alignment data. This dictionary will store all the data that we want to store and save with the face data we'll use to convert with later.

Next, it will get a list of all files in the folder. This assumes that each file is an image that we want to import any faces from. If there are any non-image files, it will fail, so it's important to keep extra files out of the folder.

Next, files in the directory are listed and checked to ensure they're files before being stored in a list.

Image preparation

The image preparation is the next step.

1.   This loads the images and gets them ready to be processed by the rest of the tools:

```
for file in tqdm(list_of_images_in_dir):
    filename, extension = os.path.splitext(file)
```

This is the start of a loop that will go over every file in the directory to process them. The tqdm library creates a nice readable status bar, including predictions for how long the process will take. This one line is enough to get the basics, but there are a lot more features that it can provide. You can see the full documentation at https://github.com/tqdm/tqdm#documentation.

2.   Next, we load the image and convert it to RGB color order:

```
image_bgr = cv2.imread(os.path.join(opt.path,file))
image_rgb = cv2.cvtColor(image_bgr, cv2.COLOR_BGR2RGB)
height, width, channels = image_rgb.shape
```

OpenCV is a library of tools for operating on images. You can find documentation for it at https://docs.opencv.org/. In this code, we use it to open images and handle various image tasks, which we'll explain as they come up.

OpenCV loads the images in a **blue, green, red** (**BGR**) color order, which is unusual, so we must convert it to **red, green, blue** (**RGB**) color order for later processing since the other libraries expect the files in that color order. If we don't convert it, all the colors will be off, and many tools will provide the wrong results.

Then we get the image's shape; this will give us the height and width of the image, as well as the number of color channels (this will be three since we loaded them as color images).

3. Next, we need to check whether we need to resize the image:

```
adjustment = opt.max_detection_size / max(height, width)
if adjustment < 1.0:
  resized_image = cv2.resize(image_rgb, None,
fx=adjustment, fy=adjustment)
else:
  resized_image = image_rgb
  adjustment = 1.0
```

When you're running images through AI models, the image's dimensions can drastically change the amount of memory used. This can cause an error if it exceeds the memory available for AI to use. In this code, we're getting a maximum size from the options, then finding what adjustment we need to make to resize the image down to the maximum size and keep track of that change so that the results from the face detector can be restored to work on the original image. This process allows us to use a smaller image for the face detection AI while still using the face from the full-sized image, giving us the best resolution and details.

Face detection

Now it's time to detect the faces within the image. This process goes through each image to detect any faces.

1. First, we check the resized image for any faces:

```
faces = face_detector.detect_from_image(resized_image)
```

This runs the face detector to find any faces in the image. The library makes this a very easy process. We send the resized image through and get back a list of all the faces found as well as their bounding boxes. These boxes are all relative to the smaller, resized image to ensure we have enough memory to handle it.

2. Next, we'll iterate over each face:

```
for idx, face in enumerate(faces):
  top,left,bottom,right = (face[0:4] /
    adjustment).astype(int)
  confidence = face[4]
  if confidence < opt.min_confidence:
    Continue
```

As input into the loop we use the built-in Python function enumerate, which gives us a count of each of the faces as it finds them. We use this number later to identify the face by number. Then, we break out the bounding box size and divide it by the adjustment size. This restores the face bounding box that was detected to match the original image instead of the smaller resized image. We store the adjusted face detection in variables that we round to integers so we can keep it lined up with the individual pixels.

Next, the confidence level of the face detection AI is used to skip any faces that fall below the confidence level, which is set up in the options.

> **Tip**
>
> The confidence level from the face detection AI varies from 0 (no face) to 1 (absolute certainty of a face). Most AI uses the range of 0 to 1 since it's easier for AI to work with a known range. If you are doing AI work and get results you don't expect, you might want to check whether it's been restricted down to the range of 0 to 1.

3. Next, we make sure the face is big enough to use:

```
face_height = bottom - top
face_width = right - left
face_size = face_height * face_width
if face_size/(height*width) < opt.min_size:
    continue
```

This code finds the face height and width, then uses that to find the overall size of the face, and compares it against a minimum face size, skipping any faces that are too small. It uses the original size of the frame to filter out faces that aren't the main focus more easily. This is set low but can be useful if you have lots of faces in a single image, such as in a crowd scene.

4. Next, we write out the bounding box as an image:

```
detected_face = image_bgr[y1:y2,x1:x2]
cv2.imwrite(os.path.join( opt.export_path,
                f"face_bbox_{filename}_{fn}.png"),
            detected_face)
```

This code creates an image containing just the parts of the image that fit inside the face bounding box of the face and saves it as a .png file into the output folder. This lets you see the face that is detected; however, it's not actually needed for any future step, so it can be removed if you don't want to save this data.

Face landmarking/aligning

The next step is to detect landmarks of the face and align them to a known position.

1. First, we will get alignments:

    ```
    landmarks = face_aligner.get_landmarks_from_image(
        image_rgb, detected_faces = [face[0:4]/adjustment])
    ```

    Here, we use the same library we used for face detection, passing the full image and the adjusted bounding boxes in order to get landmarks for the face. The library returns 68 landmark positions, which are based on specific points on the face.

2. Next, we draw a box:

    ```
    landmark_image = image_bgr.copy()
    landmark_image = cv2.rectangle(landmark_image,
        (top, left), (bottom, right), thickness=10,
        color=(0, 0, 0))
    ```

    Here, we generate a new copy of the original image in BGR color format so we can draw onto the image without damaging our original copy. We then use OpenCV to draw a rectangle for the detected face. Its thickness is set to 10 pixels and drawn in black.

3. Next, we create the landmarks we're going to use for alignment:

    ```
    right_eye = np.mean(landmarks[0][36:42],axis=0)
    left_eye = np.mean(landmarks[0][42:48],axis=0)
    nose_tip = landmarks[0][30]
    right_mouth = landmarks[0][48]
    left_mouth = landmarks[0][54]
    limited_landmarks = np.stack(
        (right_eye,
         left_eye,
         nose_tip,
         right_mouth,
         left_mouth))
    ```

    In this code, we split the 68 landmarks down to just 5. We use the average of the landmarks around each eye to find the average eye position, and then get the tip of the nose and the corners of the mouth and save them into a new array. This reduced set of landmarks helps to keep alignment consistent since the 68 landmarks contain a lot of noisy edge points.

4. Next, we will draw the landmarks onto the image:

```
landmark_image = cv2.rectangle(landmark_image,
   (x1,y1), (x2,y2), thickness=10, color=(0,0,0))
colors = [[255,0,0], # Blue
          [0,255,0], # Green
          [0,0,255], # Red
          [255,255,0], # Cyan
          [0,255,255]] # Yellow
for count, landmark in enumerate(limited_landmarks):
  landmark_adjusted = landmark.astype(int)
  landmark_image = cv2.circle(landmark_
image,  tuple(landmark_adjusted), radius=10,
  thickness=-1, color=colors[count])
cv2.imwrite(os.path.join(opt.export_path,
            f"face_landmarks_{filename}_{idx}.png",
          landmark_image)
```

Here, we define a set of colors, then use those to draw individual dots for each landmark position. We save them as a .png image file in the output folder. These images are for demonstration and debugging purposes and aren't ever used later, so you can remove them (and this save call) if you don't need those debug images.

5. Next, we define the mean face:

```
MEAN_FACE = np.array([[0.25, 0.22],
                      [0.75, 0.22],
                      [0.50, 0.51],
                      [0.26, 0.78],
                      [0.74, 0.78]])
```

Here, we define another array; this is based on the average face locations of the different landmarks. We use this in the next part to align the image. These numbers are based on where we want the face to be but could be any numbers that can be reliably detected on the face.

6. Next, we generate the transformation matrix to align the face:

```
warp_matrix = umeyama(limited_landmarks,
   MEAN_FACE * (opt.size*.6)+(opt.size*.2), True)
```

This is a bit complicated to understand, so we'll go into depth here. We use an algorithm to align two point sets created by Shinji Umeyama as implemented in the SciKit-Image library. This takes two sets of points, one a known set (in this case, MEAN_FACE, which we defined earlier) and the other an unknown set (in this case, the five landmark points from the detected face saved in limited_landmarks), and aligns them. Next, we multiply the landmarks by the size that we want the image to end up being and add a border around the face so that it is centered with some extra space around the edges.

The umeyama algorithm creates a matrix that we save as warp_matrix, which encodes the translations necessary to create an aligned face.

7.   Next, we add the landmarks and warp_matrix to the list of alignment data:

```
alignment_data[file] = {"landmark": landmarks,
                        "warp_matrix": warp_matrix}
```

8.   Finally, we create and write the aligned face image:

```
aligned_face = image_bgr.copy()
aligned_face = cv2.warpAffine(aligned_face,
  warp_matrix[:2], (opt.size,opt.size))
cv2.imwrite(os.path.join(opt.export_path,
            f"face_aligned_{filename}_{idx}.png"),
          aligned_face)
```

Here, the code creates a new copy of the original image and then uses the warpAffine function from the OpenCV library to apply the warp_matrix that was generated by the umeyama algorithm. The matrix includes all the information – translation (moving the image side to side or up and down), scaling (resizing to fit), and rotation – to align the face with the pre-defined landmarks. Finally, it saves that newly aligned image as a file.

> **Tip**
> While OpenCV does all image processes in BGR color order, it's fine to do any tasks that don't depend on the color order, such as this cv2.warpAffine() step here. If you do ignore the color order, you must be careful since it can get easy to forget which color order you are using, leading to complicated bugs where the colors are all wrong. In this case, since the next step will be to write the image out as a file using cv2.imwrite(), we are fine using the BGR color order image.

9.   Next, we save the data we'll need to later reconstruct the image:

```
if file not in alignment_data.keys():
  alignment_data[file] = {"faces": list()}
```

```
alignment_data[file]['faces'].append({
   "landmark": landmarks,"warp_matrix": warp_matrix})
```

We save the landmarks and the warp matrix into a dictionary, which we'll later save as a JSON file. This information is important to save for later processing steps, so we must make sure to save it.

10. Next, we're going to create a mask image:

```
mask_face = cv2.resize(aligned_face, (512, 512))
mask_face = torch.tensor(mask_face,
   device=device).unsqueeze(0)
mask_face = mask_face.permute(0, 3, 1, 2) / 255
if device == "cuda":
   mask_face.cuda()
segments = masker(mask_face)[0]
```

The masker is another AI that has specific requirements for the image that it is given. To meet these requirements, we must first process the face image in certain ways. The first is that the masker AI expects images that are 512x512 pixels. Since our aligned faces can be different sizes, we need to make a copy of the image that is in the expected 512x512 size.

We then convert it to a PyTorch Tensor instead of a Numpy array and then unsqueeze the tensor. This adds an additional dimension since the masker works on an array containing one or more images; even though we're only feeding it one, we still need to give it that extra dimension containing our single image to match the shape expected.

Next, the masker expects the channels to be in a different order than we have them. To do this, we permute the array into the correct order. Also, traditionally images are stored in the range of 0-255, which allows for 256 variations of each separate color, but the masker expects the image colors to be a float in the range of 0-1. We divide the range by 255 to get into the expected range.

Next, if NVIDIA support is enabled, we convert the image into a CUDA variable, which converts it into a format that can be used by the GPU as well as handle being moved to the GPU.

Finally, we run the masker AI on the image that we've prepared, saving the mask output to a new variable. The masker outputs multiple arrays of information, but only the first array is useful to us now, so we save only that one and discard all the others.

11. Next, we process the masker output and save it to an image file:

```
segments = torch.softmax(segments, dim=1)
segments = torch.nn.functional.interpolate(segments,
   size=(256, 256),
```

```
    mode="bicubic",
    align_corners=False)
mask = torch.where( torch.sum(
    segments[:,desired_segments,:,:], dim=1) > .7,
    255, 0)[0]
mask = mask.cpu().numpy()
cv2.imwrite(os.path.join(opt.export_path,
            f"face_mask_{filename}_{idx}.png"),
        mask)
```

Now that the result has been returned from the masker, we still need to process it to get something useful. First, we use softmax to convert the result from absolute to relative values. This lets us look at the mask as an array of likelihoods that each pixel belongs to a particular class instead of the raw values from the model.

Next, we use interpolate, which is a Pytorch method, to resize the data back to the original face image size. We have to do this because, like the input, the output of the masker model is 512x512. We use bicubic because it gives the smoothest results, but other options could be chosen instead.

Next, we use sum and where to **binarize** the mask. Basically, we take the mask, which is a bunch of probabilities, and turn it into a mask where the only options are 255 or 0. We also use desired_segments to remove the segments that aren't useful to us. We're using .7 as a threshold here, so if we're 70% sure that a given pixel should be in the mask, we keep it, but if it's below that 70% cutoff, we throw that pixel out.

Next, we move the data back to the CPU (if it was already on the CPU, then nothing changes) and convert it to a Numpy array.

Finally, we save the mask image as a .png file so we can use it later.

12. The last step of the entire extract process is to write the alignment data as a file:

```
with open(os.path.join(opt.export_path,
    f"face_alignments.json", "w") as alignment_file:
    alignment_file.write(
        json_tricks.dumps(alignment_data, indent=4))
```

Once each image is processed, the last step of the file is to save the file containing all the landmark data for later use. Here we use the `json_tricks` library, which has some useful functionality for writing out Numpy arrays as a JSON file. The library handles everything for writing and reading back the JSON file as Numpy arrays, so we can simply pass the full `dictionary` of arrays without manually converting them to lists or other default Python types for the Python standard library JSON to handle. For full documentation on this, please visit their documentation page at `https://json-tricks.readthedocs.io/en/latest/`.

At this point, we've extracted all the faces from a folder full of images. We've run them through multiple AI to get the data we need and formatted all that data for later use. This data is now ready for training, which will be covered in the next chapter.

Summary

Extraction is the first step of the training process. In this chapter, we examined the data that we will need in later steps as well as the process of extracting the required training data from the source images. We went hands-on with the process, using multiple AIs to detect and landmark faces and generate a mask, as well as the necessary steps to process and save that data.

The `C5-face_detection.py` file can process a directory of images. So, we covered how to convert a video into a directory of images and how to process that directory through the script. The script creates all the files you need for training and some interesting debug images that let you visualize each of the processes the detector uses to process the images. We then looked at the entire process, line by line, so that we knew exactly what was going on inside that script, learning not just what was being output, but exactly how that output was created.

After finishing the detection process, you can go through data cleaning, as talked about in *Chapter 3, Mastering Data*, to make sure your data is ready for the subject of the next chapter: training.

Exercises

1. We used pre-existing libraries for face detection, landmarking, and aligning landmarks. There are other libraries that offer similar functionality. Not all libraries work the same way, and implementing the differences is an extremely useful exercise. Try replacing the `face_alignment` library with another library for detecting faces, such as `https://github.com/timesler/facenet-pytorch` or `https://github.com/serengil/deepface`. Open source has lots of useful libraries but learning the differences and when to use one over another can be difficult, and converting between them can be a useful practice.

2. We used 2D landmarks for alignment in this chapter, but there may be a need for 3D landmarks instead. Try replacing the following:

    ```
    face_aligner = FaceAlignment(LandmarksType._2D,
      device=device, verbose=False)
    ```

    with:

    ```
    face_aligner = FaceAlignment(LandmarksType._3D,
      device=device, verbose=False)
    ```

    and adjust the rest of the process accordingly. You will also need to modify MEAN_FACE to account for the third dimension.

    What other problems do 3D landmarks include? What do you gain by using them?

3. In deepfakes, we're most interested in faces, so this process uses techniques specific to faces. Imagine what you'd need to do to extract images of different objects. Watches, hats, or sunglasses, for example. The repo at `https://github.com/ultralytics/yolov5` has a pre-trained model that can detect hundreds of different objects. Try extracting a different object instead of faces. Think in particular about how to do an alignment: can you utilize edge detection or color patterns to find points to which you can align?

4. Umeyama's method treats every point that it aligns with equal importance, but what happens if you try to align with all 68 landmarks instead of just the 5? What about 2 points? Can you find a better method? A faster method? A more accurate one? Try modifying the script to output all 68 landmark points in the `face_landmarks .png` file so you can visualize the process.

6

Training a Deepfake Model

Training a deepfake model is the most important part of creating a deepfake. It is where the AI actually learns about the faces from your data and where the most interesting neural network operations take place.

In this chapter, we'll look into the training code and the code that actually creates the AI models. We'll look at the submodules of the neural network and how they're put together to create a complete neural network. Then we'll go over everything needed to train the network and end up with a model ready to swap two faces.

We'll cover the following topics in this chapter:

- Understanding convolutional layers
- Getting hands-on with AI
- Exploring the training code

By the end of this chapter, we'll have designed our neural networks and built a training pipeline capable of teaching them to swap faces.

Technical requirements

To run any of the code in this chapter, we recommend downloading our official repository at `https://github.com/PacktPublishing/Exploring-Deepfakes` and following the instructions for setting up an Anaconda environment with all of the required libraries.

In order to train, you must have two sets of extracted faces. You'll feed both sets into the model and it will learn both faces separately. It's important that you get sufficient data for both faces and that there be a good variety. If you're in doubt, please check *Chapter 3, Mastering Data*, for advice on getting the best data.

Understanding convolutional layers

In this chapter, we'll finally get into the meat of the neural networks behind deepfakes. A big part of how networks such as these work is a technique called convolutional layers. These layers are extremely important in effectively working with image data and form an important cornerstone of most neural networks.

A **convolution** is an operation that changes the shape of an object. In the case of neural networks, we use **convolutional layers**, which iterate a convolution over a matrix and create a new (generally smaller) output matrix. Convolutions are a way to reduce an image in size while simultaneously searching for patterns. The more convolutional layers you stack, the more complicated the patterns that can be encoded from the original image.

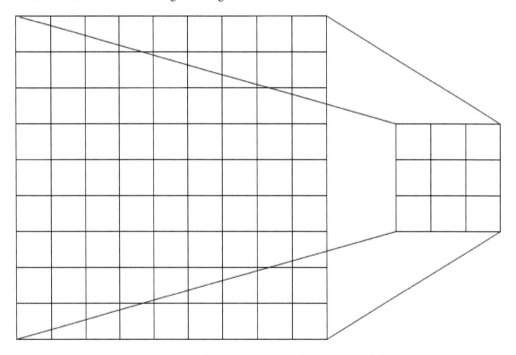

Figure 6.1 – An example of a convolution downscaling a full image

There are several details that define a convolutional layer. The first is dimensionality. In our case, we're using 2D convolutions, which work in 2D space. This means that the convolution works on the x and y axes for each of the channels. This means for the first convolution, each color channel is processed separately.

Next is the **kernel**, which defines how big an area each convolution takes into account. The amount of kernels going across affects the output as well. For example, if you had a matrix of 3x9 and kernel size of 3x3, you'd get a 1x3 matrix output.

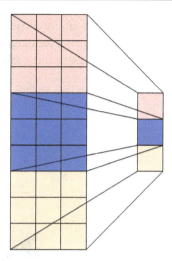

Figure 6.2 – An example of a 3x3 convolutional process turning a 3x9 into a 1x3 matrix output

Next is **stride** which defines how big a step each iteration of the convolution takes as it travels the matrix. A stride of 2, for example, would make our 3x3 kernel overlap by a single **entry** of the matrix. Stride is duplicated in every dimension, so if you extended the example input matrix to the left or right, you'd also get overlap in that direction.

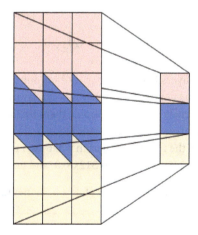

Figure 6.3 – An example of a stride smaller than the kernel size causing
the elements in between to be shared on the output

Last is **padding** which is added to the input and defines how much outside the original image the convolution should pretend is there. If we added a padding of 1 to an input matrix of 1x7, it would act like a 3x9, and we'd still get a 1x3 output. This is useful for controlling the precise size of your output matrix or for ensuring that every pixel gets covered by the center of a kernel. It's also necessary for any situation where the kernel is larger than the input matrix in any direction – though in practice, this situation is extremely rare. In our example, the empty entries are filled with 0s. However, by using `padding_mode` you can specify different types of padding, such as reflect, which will make the padding equal to the entry that it mirrors along the padding axis like a mirror at the edge of the input matrix, reflecting each entry back.

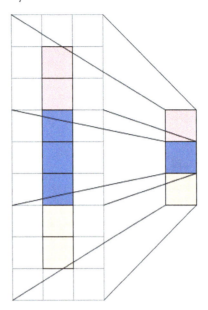

Figure 6.4 – Example of padding a 1x7 into a 3x9 before convolution into a 1x3

We stack multiple convolutional layers because we're looking for bigger patterns. To find all the appropriate patterns, we increase the depth of the convolutional layers as we add them to the tower. We start with a convolution 128 kernels deep, then double them to 256, 512, and finally 1,024. Each layer also has a kernel size of 5, a stride of 2, and a padding of 2. This effectively shrinks the width and height by half of each layer. So, the first layer takes in a 3x64x64 image and outputs a 128x32x32 matrix. The next layers turn that to 256x16x16, then 512x8x8, and finally 1024x4x4.

Next, we'll finally get into the code at the heart of deepfakes – the neural network itself.

> **Tip**
>
> It can be confusing to track how a convolution layer will change a matrix's size. The equation to calculate the output matrix's size is actually quite simple but non-intuitive: `(input_size+2*padding-stride+1)/2`. If you have square matrices, this calculation will match for either dimension, but if you have a non-square matrix, you'll have to calculate this for both dimensions separately.

Getting hands-on with AI

The first code we'll examine here is the actual model itself. This code defines the neural network and how it's structured, as well as how it's called. All of this is stored in the `lib/models.py` library file.

First, we load any libraries we're using:

```
import torch
from torch import nn
```

In this case, we only import PyTorch and its nn submodule. This is because we only include the model code in this file and any other libraries will be called in the file that uses those functions.

Defining our upscaler

One of the most important parts of our model is the upscaling layers. Because this is used multiple times in both the encoder and decoder, we've broken it out into its own definition, and we'll cover that here:

1. First, we define our class:

    ```
    class Upscale(nn.Module):
      """ Upscale block to double the width/height from
    depth. """
      def __init__(self, size):
        super().__init__()
    ```

    Note that, like our encoder, this inherits from `nn.Module`. This means we have to make a call to the initialization from the parent class in this class's initialization. This gives our class a lot of useful abilities from PyTorch, including the backpropagation algorithms that make neural networks work.

2. Next, we define our layers:

    ```
    self.conv = nn.Conv2d(size * 2, size * 2 * 2, kernel_
    size=3,
      padding="same")
    self.shuffle = nn.PixelShuffle(2)
    ```

    The upscaler only uses two layers. The first is a convolutional layer, which has an input size double that of the initialization function. We do this because the upscale class takes in an output size, and since it increases the width and height by halving the depth, it needs the input depth to be twice the output. In this case, padding is same instead of a number. This is a special way to make the nn.Conv2d layer output a matrix with the same width and height as the input. For a kernel size of 3 this creates a padding of 1.

The nn.PixelShuffle is a layer that takes an input matrix and, by moving the entries around, takes depth layers and converts them into width and height. Together with the earlier convolutional layer, this effectively "upscales" the image in a learnable and efficient way. We pass 2 since we want it to double the width and height. Other numbers can be used for different scaling factors but would require adjustments of the convolutional layers and the models that call the class.

3. Finally, we have our forward function:

```
def forward(self, x):
    """ Upscale forward pass """
    x = self.conv(x)
    x = self.shuffle(x)
    return x
```

This forward function simply takes the input, then runs it through the convolutional and PixelShuffle layers and returns the result.

Creating the encoder

Let's create the encoder next:

1. First, we declare the encoder class:

```
class OriginalEncoder(nn.Module):
    """ Face swapping encoder

    Shared to create encodings for both the faces
    """
```

Here we've defined and provided a short comment on the encoder. We declare it as a child class of the nn.Module class. This gives our class a lot of useful abilities from PyTorch, including the backpropagation algorithms that make neural networks work.

> **Author's note**
>
> In this book, we've included only the basic, original model. This was the first deepfake model and has been surpassed in pretty much every way, but it's easy to understand, so it works well for this book. If you'd like to explore other models, we recommend that you check out Faceswap at https://Faceswap.dev, which is constantly updated with the newest models.

2. Next, we'll define the initialization function:

```
def __init__(self):
    super().__init__()
```

This function is the one that actually builds the neural network's layers. Each layer is defined in this function so that PyTorch can automatically handle the details of the weights. We also call the __init__ function from the parent class to prepare any variables or functionality that is necessary.

3. Next, we'll start defining our activation function:

```
self.activation = nn.LeakyReLU(.1)
```

We use LeakyReLU or **Leaky Rectified Linear Unit** as an **activation function** for our model. An activation function takes the output of a layer and brings it into a standardized range.

What a Leaky Rectified Linear Unit *is* is pretty easy to understand if you break down the words from last to first. *Unit*, in this case, means the same as function; it takes an input and provides an output. *Linear* means a line, one that doesn't change directions as it moves; in this case, it's a 1:1, where the output matches the input (an input of 1 leads to an output of 1, an input of 2 leads to an output of 2, and so on). *Rectified* just means it has been made positive, so negative numbers become 0. *Leaky* actually makes that last sentence a bit of a lie. It's been found that neural networks really don't work very well when the entire negative space becomes 0. So leaky here means that negative numbers get scaled to a range barely below 0.

We use 0.1 here so that any numbers below 0 get multiplied by 0.1, scaling them smaller by 10 times. Many different values can be used here, and various projects make different decisions. Standard values typically sit somewhere in the range of 0.005 to 0.2.

4. Next, we'll define our convolution tower:

```
self.conv_tower = nn.Sequential(
    nn.Conv2d(3, 128, kernel_size=5, stride=2, padding=2),
    self.activation,
    nn.Conv2d(128, 256, kernel_size=5, stride=2,
padding=2),
    self.activation,
    nn.Conv2d(256, 512, kernel_size=5, stride=2,
padding=2),
    self.activation,
    nn.Conv2d(512, 1024, kernel_size=5, stride=2,
padding=2),
    self.activation)
```

The convolution tower is exactly what it sounds like, a stack of convolution layers. After each of the convolution layers, we include an activation function. This is helpful to ensure that the model stays on track and makes the convolutions more effective. The activation is identical in each case and doesn't do any "learning" but just works like a function, so we don't need to make separate layers for each one and can use the same activation function we already initialized.

We use nn.Sequential here to combine the stack of layers into a single layer. The sequential layer is actually a very powerful tool in PyTorch, allowing you to make simple neural networks without having to write a whole class for the model. We use it here to combine all the convolutional layers since the input in one end goes all the way through in every case. This makes it easier to use later in our forward function. But a sequential model runs each of its constituent layers in sequence and can't handle conditional if statements or functions that aren't written for PyTorch.

5. Next, we'll define a flatten layer:

    ```
    self.flatten = nn.Flatten()
    ```

    A flatten layer does exactly what it sounds like; it flattens a previous layer to just one axis. This is used in the forward pass to turn the 1024x4x4 matrix that comes out of the convolution tower into a 4,096-element wide single-dimension layer.

6. Next, we'll define our dense layers:

    ```
    self.dense1 = nn.Linear(4 * 4 * 1024, 1024)
    self.dense2 = nn.Linear(1024, 4 * 4 * 1024)
    ```

    Dense layers are called dense because they're fully connected. Unlike convolutional layers, every single entry in the matrix is connected to every single input of the previous layer. Dense layers were the original neural network layer types and are very powerful, but they're also very memory intensive. In fact, these two layers account for most of the memory of the entire deepfake model!

    We generate two separate dense layers. The first layer takes in an input of 4,096 entries wide and outputs a 1,024-wide output. This is the **bottleneck** of the model: the part of the model that has the least amount of data, which then needs to be rebuilt. The second layer takes a 1024 one-dimensional matrix input and outputs a matrix with one dimension of 4,096. This is the first layer that starts rebuilding a face from encoded details.

7. The last initialization step is to define our first upscale layer:

    ```
    self.upscale = Upscale(512)
    ```

    This layer is our first upscaler. This layer will take a 1024x4x4 matrix and upscale it back to a 512x8x8 matrix. All other upscalers will exist in the decoder. This one was originally put in the encoder, probably as a memory-saving attempt since the first upscale was unlikely to need to match a particular person at all, as it only had the most general of face patterns.

The upscale layer is given an output size of 512. This means that the output will be 512 deep but does not define the width or height. These come naturally from the input, with each call to upscale doubling the width and height.

8. Next, we'll go over our forward function:

```
def forward(self, x):
    """ Encoder forward pass """
```

The forward function is what actually applies the network to a given matrix. This is used both for training and for inference of the trained model.

9. First, we get the batch size:

```
batch_size = x.shape[0]
```

We need the batch size that we started with later in the process, so we save it here immediately.

10. Finally, we run the data through the whole model:

```
x = self.conv_tower(x)
x = self.flatten(x)
x = self.dense1(x)
x = self.dense2(x)
x = torch.reshape(x, [batch_size, 1024, 4, 4])
x = self.upscale(x)
x = self.activation(x)
return x
```

In this code, we run the input matrix through each layer in turn. The only new surprise here is the `torch.reshape` call after the final `dense`, which is effectively the opposite of the `flatten` call from right before the first `dense`. It takes the 4096-wide matrix and changes the shape so that it's a 1024x4x4 matrix again.

We then run the data through the upscale layer and then the activation function before we return the result.

Building the decoders

The decoder is responsible for taking the encoded face data and re-creating a face as accurately as it can. To do this, it will iterate over thousands or even millions of faces to get better at turning encodings into faces. At the same time, the encoder will be getting better at encoding faces.

We used the plural *decoders* here, but this code only actually defines a single decoder. That's because the training code creates two copies of this decoder class.

1. First, we define and initialize our model:

```
class OriginalDecoder(nn.Module):
  """ Face swapping decoder
  An instance for each face to decode the shared
encodings.
  """
  def __init__(self):
    super().__init__()
```

This code, just like the encoder and upscaler, is an instance of nn.Module and needs an initialization function that also calls the parent's initializer.

2. Next, we define our activation function:

```
self.activation = nn.LeakyReLU(.1)
```

Just like our encoder's activation, we use LeakeReLu with a negative scaling of 0.1.

3. Next, we define our upscaling tower:

```
self.upscale_tower = nn.Sequential(Upscale(256),
  self.activation,
  Upscale(128),
  self.activation,
  Upscale(64),
  self.activation)
```

The upscale tower is much like the convolution tower of the encoder but uses upscale blocks instead of shrinking convolutions. Because there was one upscaler in the encoder, we actually have one fewer upscales in this decoder. Just like the convolution tower, there are also activation functions after each upscale to keep the range trending positive.

4. Next, we define our output layer:

```
self.output = nn.Conv2d(64, 3, 5, padding="same")
```

The output layer is special. While each of the previous layers' outputs was half the depth of the previous layer's, this one takes the 64-deep output from the convolution layer and converts it back to a three-channel image. There is nothing special about the three-channel dimension, but due to how the training process works, each is correlated to one of the color channels of the training image.

5. Now, we define the forward function of the decoder:

```
def forward(self, x):
    """ Decoder forward pass """
    x = self.upscale_tower(x)
    x = self.output(x)
    x = torch.sigmoid(x)
    return x
```

This forward function is familiar, being very similar to those in the encoder and the upscale layer. The major difference here is that after we pass the input through the upscale tower and the output layer, we use a `torch.sigmoid` layer. This is another type of activation layer.

Sigmoid works differently from LeakyReLu in that it is not linear. Instead, it computes the logistic sigmoid of the input. This is an s-shaped output where negative inputs approach 0, and positive inputs approach 1 with a 0 input coming out as 0.5. The precise equation is 1/ (1*e^-input). This basically puts the results between 0 and 1 with extremes being more compressed, which matches how the multiplication of high numbers leads to higher numbers faster. This effectively turns the output of the model into a range of 0-1, which we can easily turn into an image.

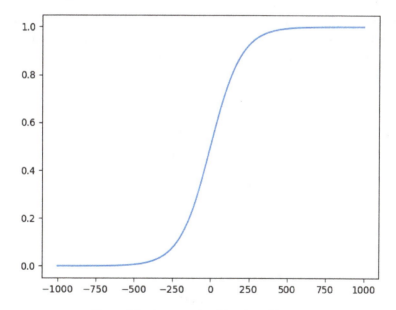

Figure 6.5 – An example of the sigmoid curve

Next, we'll examine the training code.

Exploring the training code

Now that we have defined our models, we can go ahead with the process of training a neural network on our data. This is the part where we actually have AI learn the different faces so that it can later swap between them.

1. First, we import our libraries:

```
from glob import glob
import os
import random
from argparse import ArgumentParser
import cv2
import numpy as np
from tqdm import tqdm
import torch


from lib.models import OriginalEncoder, OriginalDecoder
```

Like all Python programs, we import our libraries. We also import our encoder and decoders from our model file. This loads the AI model code from earlier in this chapter and lets us use those to define our models in this code. Python really makes it easy to import code we've already written, as every Python file can be called directly or imported into another file.

Note that Python uses a strange syntax for folder paths. Python treats this syntax exactly the same as a module, so you use a period to tell it to look in a folder and then give it the file you want. In this case, we're pulling the `OriginalEncoder` and `OriginalDecoder` classes from the `models.py` file located in the `lib` folder.

2. Next, we define our arguments and call our main function:

```
If __name__ == "__main__":
    # Train a deepfake model from two folders of face
images.
    #    Example CLI:
    #    ------------
    #    python c6-train.py "C:/media/face1"
                            "C:/media/face2"
```

3. Next, we define our arguments:

```
parser = ArgumentParser()
parser.add_argument("patha",
  help="folder of images of face a")
```

```
parser.add_argument("pathb",
  help="folder of images of face b")
parser.add_argument("--cpu",
  action="store_true",
  help="Force CPU usage")
parser.add_argument("--batchsize",
  type=int, default=16,
  help="Number of images to include in a batch")
parser.add_argument("--iterations", type=int,
default=100000,
  help="Number of iterations to process before stopping")
parser.add_argument("--learning-rate",
  type=float, default=.000001,
  help="Number of images to include in a batch")
parser.add_argument("--save_freq",
  type=int, default=1000,
  help="Number of iterations to save between")
parser.add_argument("--out_path",
  default="model/",
  help="folder to place models")
```

Here we define our arguments. These give us the ability to change our settings, files, or details without having to modify the source code directly.

4.   Then, we parse all the arguments and call our main function:

```
opt = parser.parse_args()
main(opt)
```

We parse our arguments and pass them into our main function. The main function will handle all the training processes, and we need to give it all the arguments.

5.   Next, we start our main function:

```
def main(opt):
  """ Train a deepfake model from two folders of face
images.
  """
  device = "cuda" if torch.cuda.is_available() and not
opt.cpu else "cpu"
os.makedirs(opt.out_path, exist_ok=True)
```

Here we start our main function and check whether we're supposed to use cuda. If so, we enable cuda so that we can use the **graphics processing unit** (**GPU**) to accelerate training. Then we create our export folder if that isn't already created. This is where we'll save copies of our models and any training previews we generate later.

> **Author's note**
>
> While it's possible to run the other parts of the process without a GPU, training is far more intensive, and running a training session on a **central processing unit** (**CPU**) will take a very large amount of time. Because of this, it's recommended that at least this part be run with a GPU. If you don't have one locally, you can rent one at any number of online services.

Creating our models

Here we'll create our neural models and fill them with weights:

1. First, we'll create instances of our previous models:

    ```
    encoder = OriginalEncoder()
    decodera = OriginalDecoder()
    decoderb = OriginalDecoder()
    ```

    In this chunk of code, we create our AI models. We create one instance of the encoder and two separate decoders. We call them a and b here, but that's entirely an arbitrary choice with no effect on the results. By default, we assume that you want to put the second face onto the first so in the case of this code, we'd be putting the face from b onto the frame from a.

2. Next, we load any previously saved models:

    ```
    if os.path.exists(os.path.join(opt.out_path, "encoder.
    pth")):
      encoder.load_state_dict( torch.load(
        os.path.join(opt.out_path, "encoder.pth")).state_
    dict())
      decodera.load_state_dict( torch.load(
        os.path.join(opt.out_path, "decodera.pth")).state_
    dict())
      decoderb.load_state_dict( torch.load(
        os.path.join(opt.out_path, "decoderb.pth")).state_
    dict())
    ```

    Here we check whether any models already exist in the given output folder. If they do, we load those model weights into the models we instantiated in the last section. To do this, we have PyTorch load the weights from the disk and then assign the weights to the model's state dictionary. This lets PyTorch load the weights into the model and get it ready for training.

If there are no weights, then we skip this step. This means that the models will be initialized with random weights, ready to start a new training session. This lets you get started easily without having to generate any random weights yourself.

3. Next, we get a list of the images to train with:

```
imagesa = glob(os.path.join(opt.patha, "face_aligned_*.
png"))
imagesb = glob(os.path.join(opt.pathb, "face_aligned_*.
png"))
```

This chunk gets a list of all the images from the folders to train. We load only the aligned face images from the folders since we can create the filenames for the other images from the filenames for the aligned images.

4. Next, we create the tensors for the images and the masks:

```
img_tensora = torch.zeros([opt.batchsize, 3, 64, 64])
img_tensorb = torch.zeros([opt.batchsize, 3, 64, 64])
mask_tensora = torch.zeros([opt.batchsize, 1, 64, 64])
mask_tensorb = torch.zeros([opt.batchsize, 1, 64, 64])
```

Here we create the tensors that will hold the images we will use for training. For the image tensors, we create a tensor that is 64x64 pixels wide with 3 channels to handle the red, green, and blue color channels. We also add a **batch size** dimension to the tensor so we can store that many images at once. A batch is simply how many images we'll process at the same time. Larger batch sizes help the training process run more efficiently as we're able to benefit from hardware that can do multiple tasks simultaneously as well as benefit from PyTorch grouping the tasks in the best order.

Author's note

Larger batch sizes lead to more efficiencies in the training, so why don't we set the batch size to 256 or 1024 instead of defaulting to 16? It's because batch size is not a magic bullet. First, larger batches take more memory as the system must store every item of the batch at the same time. With large models, this can be prohibitive. Additionally, there is a side effect to large batch sizes. It's the classic "forest for the trees," meaning that larger batch sizes can help generalize over large sets of data but perform worse at learning specific details. So, picking the ideal batch size can be as important a question as anything else. A good rule of thumb for deepfakes is to keep batch size in double digits with 100+ tending to be too big and <10 to be avoided unless you have specific plans.

---

5. Next, we define and set up our optimizers and loss function:

```
encoder_optimizer = torch.optim.Adam(
    encoder.parameters(), lr=opt.learning_rate/2)
```

```
decodera_optimizer = torch.optim.Adam(
  decodera.parameters(), lr=opt.learning_rate)
decoderb_optimizer = torch.optim.Adam(
  decoderb.parameters(), lr=opt.learning_rate)
loss_function = torch.nn.MSELoss()
```

Optimizers are the part of PyTorch responsible for the most important part of training: backpropagation. This process is what changes the weights of the model and allows AI to "learn" and get better at re-creating the images we're using for training. It's responsible for more than just changing the weights but actually calculates how much to change them as well.

In this case, we're using the `torch.optim.Adam` optimizer. This is part of a family of optimizers proven to be very effective and flexible. We use it here because it's what the original deepfake model used, but it's still one of the most reliable and useful optimizers even today.

We pass each model the **learning rate** from our options. The learning rate is basically a scaling value of how much the optimizer should change the weights. Higher numbers change the weights more, which can lead to faster training at the cost of difficulty in fine-tuning since the changes being made are large. Lower learning rates can get better accuracy but cause training to take longer by being slower. We cut the learning rate of the encoder by half because we will actually be training it twice as often since it is being used to encode both faces.

The last thing we do here is to define our **loss function**. This is the function that is responsible for providing a "score" of how well AI did its job. In this case, we use the `torch.nn.MSEloss` provided by PyTorch. This is a loss called **mean squared error** (**MSE**). Let's look at this word by word again.

Error in mathematics is how far off the function is from a perfectly correct result. In our case, we are re-creating a face, so the loss function will compare the generated face to the original face and count how far off each pixel is from the correct answer. This gives a nice easy number for each pixel. Looking at each pixel is a bit too difficult, so next, our loss will take the average (mean) of all the pixels together. This gives a single number of how far off the AI was as a whole. Finally, that number is squared. This makes large differences stand out even more and has been shown to help the model reach a good result faster.

There are other loss functions such as **mean absolute error** (**MAE**), which gets rid of the squaring in the MSE, or **structural similarity**, which uses the similarity of the structure as a measure. In fact, **generative adversarial networks** (**GANs**), which are a buzzword of the machine learning field, simply replace the static loss of an auto-encoder with another model that provides a trainable loss function and pits the two models against each other in a competition of which model can do their job better.

6. Next, we move everything to the GPU if enabled:

```
if device == "cuda":
  encoder = encoder.cuda()
```

```
decodera = decodera.cuda()
decoderb = decoderb.cuda()
img_tensora = img_tensora.cuda()
img_tensorb = img_tensorb.cuda()
mask_tensora = mask_tensora.cuda()
mask_tensorb = mask_tensorb.cuda()
```

If cuda was enabled earlier, we need to move the models and the variables to the GPU so we can process them. So here, we check whether cuda was enabled, and if so, we move each of them to the GPU.

Looping over the training

In order to train the model, we need to loop over all the data. We call this the training loop.

1. First, we create a progress bar and start the training loop:

```
pbar = tqdm(range(opt.iterations))


for iteration in pbar:
```

We use tqdm again for a progress bar. Here we pass a range of how many iterations we want to tqdm so it can update our progress bar automatically and assign the progress bar to a variable. We then start our loop from that variable to provide more information in the progress bar by calling functions that tqdm exposes in the variable.

2. Next, we load a random set of images:

```
images = random.sample(imagesa, opt.batchsize)
for imgnum, imagefile in enumerate(images):
    img = cv2.imread(imagefile)
    img = cv2.resize(image, (64, 64))
    mask = cv2.imread(imagefile.replace("aligned", "mask"),
0)
    mask = cv2.resize(mask, (64, 64))
    if np.random.rand() > .5:
        image = cv2.flip(img, 1)
        mask = cv2.flip(mask, 1)
    img_tensor = torch.tensor(img[...,::-1]/255).
permute(2,0,1)
    mask_tensor = torch.where(torch.tensor(mask) > 200, 1,
0)
```

```
if device == "cuda":
  img_tensor = img_tensor.cuda()
  mask_tensor = mask_tensor.cuda()
img_tensora[imgnum] = img_tensor
mask_tensora[imgnum] = mask_tensor
```

This chunk loads a set of images from the a set for the model to train with. To do this, we get a random sample the same size as our batch size from the list of files we generated earlier.

Next, we go over a loop for each of those images. The loop first reads in the face image and resizes it down to 64x64. Then it does the same for the mask image by replacing the "aligned" word in the filename with "mask", which matches the mask filenames. The mask is also resized to match the training image.

Next, we randomly get a 50% chance to flip the images horizontally. This is an extremely common way to get more variety out of the dataset. Since faces are generally pretty symmetrical, we can usually flip them. We use a 50% chance here, which gives us an equal chance of the image being flipped or not. Since we have a mask, we have to flip it, too, if we flip the image.

Next, we convert both the image and masks from image arrays into tensors. To do this to the image, we convert from **blue, green, red (BGR)** to **red, green, blue (RGB)**, divide by 255, and move the channels to the first dimension. This is brought into a single-liner, but it is the same process as we used in the last chapter, where we were finding faces. Since BGR and RGB are just reversed, we can change BGR to RGB by just reversing the channels by indexing with [...,::-1]. This can also be done again to get it back to the BGR order (which we'll do later). The mask is simpler since we don't care about color data for it, so we just see if the pixel data is greater than 200; if it is, we put 1 into the tensor; if not, we put 0.

Next, we check whether cuda is enabled; if it is, we move the tensors we just created to the GPU. This puts everything onto the same device.

Finally, we move the image into the tensor we'll be using to train the model. This lets us batch the images together for efficiency.

3. Then, we do the same for the other set of images:

```
images = random.sample(imagesb, opt.batchsize)
for imgnum, imagefile in enumerate(images):
  img = cv2.imread(imagefile)
  img = cv2.resize(image, (64, 64))
  mask = cv2.imread(imagefile.replace("aligned", "mask"),
0)
  mask = cv2.resize(mask, (64, 64))
  if np.random.rand() > .5:
    image = cv2.flip(img, 1)
```

```
    mask = cv2.flip(mask, 1)
  img_tensor = torch.tensor(img[...,::-1]/255).
permute(2,0,1)
  mask_tensor = torch.where(torch.tensor(mask) > 200, 1,
0)
  if device == "cuda":
    img_tensor = img_tensor.cuda()
    mask_tensor = mask_tensor.cuda()
  img_tensorb[imgnum] = img_tensor
  mask_tensorb[imgnum] = mask_tensor
```

This code is identical to the last code but is repeated for the b set of images. This creates our second set of images ready to train.

Teaching the network

Now it's time to actually perform the steps that train the network:

1. First, we clear the optimizers:

    ```
    Encoder_optimizer.zero_grad()
    decodera_optimizer.zero_grad()
    ```

    PyTorch is very flexible in how it lets you build your models and training process. Because of this, we need to tell PyTorch to clear the **gradients** each time we start a training iteration. Gradients are a log of the results of each image going through each layer. These are used to know how much each weight needs to be modified during the backpropagation step. In this chunk we clear these gradients for the encoder and the a side decoder.

2. Then, we pass the images through the encoder and decoder:

    ```
    Outa = decodera(encoder(img_tensora))
    ```

    This chunk sends the image tensors we created earlier through the encoder and then through the decoder and stores the results for comparison. This is actually the AI model, and we give it a tensor of images and get a tensor of images back out.

3. Next, we calculate our loss and send it through the optimizers:

    ```
    Lossa = loss_function(
      outa * mask_tensora, img_tensora * mask_tensora)
    lossa.backward()
    encoder_optimizer.step()
    decodera_optimizer.step()
    ```

This chunk does the rest of the training, calculating the loss and then having the optimizers perform backpropagation on the models to update the weights. We start by passing the output images and the original images to the loss function.

To apply the masks, we multiply the images by the masks. We do this here instead of before we pass the images to the model because we might not have the best masks, and it's better to train the neural network on the whole image and apply the mask later.

Next, we call `backward` on the loss variable. We can do this because the variable is actually still a tensor, and tensors keep track of all actions that happened to them while in training mode. This lets the loss be carried back over all the steps back to the original image.

The last step is to call the `step` function of our optimizers. This goes back over the model weights, updating them so that the next iteration should be closer to the correct results.

4. Next, we do the same thing, but for the b decoder instead:

    ```
    encoder_optimizer.zero_grad()
    decoderb_optimizer.zero_grad()

    outb = decoderb(encoder(img_tensorb))

    lossb = loss_function(
        outb * mask_tensorb, img_tensorb * mask_tensorb)
    lossb.backward()
    encoder_optimizer.step()
    decoderb_optimizer.step()
    ```

    We go through the same process again with the b images and decoder. Remember that we're using the same encoder for both models, so it actually gets trained a second time along with the b decoder. This is a key part of how deepfakes can swap faces. The two decoders share a single encoder, which eventually gives both the decoders the information to re-create their individual faces.

5. Next, we update the progress bar with information about this iteration's loss of data:

    ```
    pbar.set_description(f"A: {lossa.detach().cpu().
    numpy():.6f} "
        f"B: {lossb.detach().cpu().numpy():.6f}")
    ```

    Since the loss function outputs a number for the optimizers, we can also display this number for the user. Sometimes loss is used by deepfakers as an estimate of how finished the model is training. Unfortunately, this cannot actually measure how good a model is at converting one face into another; it only scores how good it is at re-creating the same face it was given. For this reason, it's an imperfect measure and shouldn't be relied on. Instead, we recommend that the previews we'll be generating later be used for this purpose.

Saving results

Finally, we will save our results:

1. First, we'll check to see whether we should trigger a save:

```
if iteration % opt.save_freq == 0:
  with torch.no_grad():
    outa = decodera(encoder(img_tensora[:1]))
    outb = decoderb(encoder(img_tensorb[:1]))
    swapa = decoderb(encoder(img_tensora[:1]))
    swapb = decodera(encoder(img_tensorb[:1]))
```

We want to save regularly – an iteration may take less than a second, but a save could take several seconds of writing to disk. Because of this, we don't want to save every iteration; instead, we want to trigger a save on a regular basis after a set number of iterations. Different computers will run at different speeds, so we let you set the save frequency with an argument.

One thing we want to save along with the current weights is a preview image so we can get a good idea of how the model is doing at each save state. For this reason, we'll be using the neural networks, but we don't want to train while we're doing this step. That's the exact reason that `torch` has the `torch.no_grad` context. By calling our model from inside this context, we won't be training and just getting the results from the network.

We call each decoder with samples of images from both faces. This lets us compare the re-created faces along with the generated swaps. Since we only want a preview image, we can throw out all but the first image to be used as a sample of the current stage of training.

2. Next, we create the sample image:

```
example = np.concatenate([
  img_tensora[0].permute(1, 2, 0).detach().cpu().numpy(),
  outa[0].permute(1, 2, 0).detach().cpu().numpy(),
  swapa[0].permute(1, 2, 0).detach().cpu().numpy(),
  img_tensorb[0].permute(1, 2, 0).detach().cpu().numpy(),
  outb[0].permute(1, 2, 0).detach().cpu().numpy(),
  swapb[0].permute(1, 2, 0).detach().cpu().numpy()
  ],axis=1)
```

We need to create our sample image from all the parts. To do this, we need to convert all the image tensors into a single image. We use `np.concatenate` to join them all into a single array along the width axis. To do this, we need to get them all into image order and convert them to NumPy arrays. The first thing we do is drop the batch dimension by selecting the first one. Then we use `permute` to reorder each tensor, so the channels are last. Then we use `detach` to remove any gradients from the tensors. We can then use `cpu` to bring the weights back onto the CPU. Finally, we use `numpy` to finish converting them into NumPy arrays.

3. Next, we write the preview image:

```
cv2.imwrite(
    os.path.join(opt.out_path, f"preview_{iteration}.png"),
    example[...,::-1]*255)
```

This chunk uses `cv2.imwrite` from OpenCV to write out the preview image as a PNG file. We put it in the output path and give it a name based on what iteration this is. This lets us save each iteration's preview together and track the progress of the network over time. To actually write out a usable image, we have to convert the color space back to the BGR that OpenCV expects, and then we multiply by `255` to get a result that fits into the integer space.

4. Next, we save the weights to a file:

```
torch.save(encoder,
    os.path.join(opt.out_path, "encoder.pth"))
torch.save(decodera,
    os.path.join(opt.out_path, "decodera.pth"))
torch.save(decoderb,
    os.path.join(opt.out_path, "decoderb.pth"))
```

Here we save the weights for our encoder and both decoders by calling `torch.save` with the output path and the filenames we want to use to save the weights. PyTorch automatically saves them to the file in its native `pth` format.

5. Finally, we save again:

```
torch.save(encoder,
    os.path.join(opt.out_path, "encoder.pth"))
torch.save(decodera,
    os.path.join(opt.out_path, "decodera.pth"))
torch.save(decoderb,
    os.path.join(opt.out_path, "decoderb.pth"))
```

For our last step, we repeat the save code. But, actually, this is done outside of the training loop. This is here just in case the training loop ends on an iteration number that doesn't trigger the save there. This way, the model is definitely saved at least once, no matter what arguments are chosen by the user.

Author's note

This may seem obvious to anyone who has spent any time coding, but it's worth repeating. You should always consider how choices made in the development process might lead to bad or unexpected outcomes and account for those in your design. It's obviously impossible to consider everything, but sometimes even something as simple as duplicating a save right before exiting, "just in case," can save someone's day (or in the case of a very long training session, even more time).

Summary

In this chapter, we trained a neural network to swap faces. To do this, we had to explore what convolutional layers are and then build a foundational upscaler layer. Then we built the three networks. We built the encoder, then two decoders. Finally, we trained the model itself, including loading and preparing images, and made sure we saved previews and the final weights.

First, we built the models of the neural networks that we were going to train to perform the face-swapping process. This was broken down into the upscaler, the shared encoder, and the two decoders. The upscaler is used to increase the size of the image by turning depth into a larger image. The encoder is used to encode the face image down into a smaller encoded space that we then pass to the decoders, which are responsible for re-creating the original image. We also looked at activation layers to understand why they're helpful.

Next, we covered the training code. We created instances of the network models, loaded weights into the models, and put them on the GPU if one was available. We explored optimizers and loss functions to understand the roles that they play in the training process. We loaded and processed images so that they were ready to go through the model to assist in training. Then we covered the training itself, including how to get the loss and apply it to the model using the optimizers. Finally, we saved preview images and the model weights themselves so we could load them again.

In the next chapter, we'll take our trained model and use it to "convert" a video, swapping the faces.

Exercises

1. Choosing learning rate is not a solved problem. There is no one "right" learning rate. What happens if you make the learning rate 10 times bigger? 10 times smaller? What if you start with a large learning rate and then reduce it after some training?

2. We used the same loss function and optimizer as the original code, which was first released back in 2018, but there are a lot of options now that weren't available then. Try replacing the loss function with others from PyTorch's extensive collection (`https://pytorch.org/docs/stable/nn.html#loss-functions`). Some of them will work without any change, but some won't work for our situation at all. Try different ones, or even try combinations of loss functions!

3. We defined a model that downscaled from a 64x64 pixel image and re-created that same image. But with some tweaks, this same architecture can instead create a 128x128 or 256x256 pixel image. How would you make changes to the model to do this? Should you increase the number (and size) of layers in the convolutional tower and keep the bottleneck the same, increase the size of the bottleneck but keep the layers the same, or change the convolutional layer's kernel sizes and strides? It's even possible to send in a 64x64 pixel image and get out a 128x128 pixel image. All of these techniques have their advantages and drawbacks. Try each out and see how they differ.

4. We're training on just two faces, but you could potentially do more. Try modifying the training code to use three different faces instead of just two. What changes would you need to make? What modifications would you make so that you can train an arbitrary number of faces at once?

5. In the years since deepfakes were first released, there have been a lot of different models created. Faceswap has implementations for many newer and more advanced models, but they're written in Keras for Tensorflow and can't work in this PyTorch fork without being ported. Check the Faceswap models at the GitHub repo (`https://github.com/deepfakes/Faceswap/tree/master/plugins/train/model`). Compare this model to the one in `Original.py`, which implements the same model. Now use that to see how `dfaker.py` differs. The residual layer works by adding an extra convolution layer and an `add` layer, which just adds two layers together. Can you duplicate the `dfaker` model in this code base? What about the others?

7

Swapping the Face Back into the Video

In this chapter, we'll complete the deepfake process by converting the videos to swap faces using the models trained in the last chapter.

Conversion is the last step of deepfaking, and it is the part that actually puts the new face onto the existing video. This requires you to already have a video that you have fully processed through the extraction process in *Chapter 5, Extracting Faces*, and uses a trained model from *Chapter 6, Training a Deepfake Model*.

We will cover the following topics in this chapter:

- Preparing to convert video
- Getting hands-on with the convert code
- Creating the video from images

Technical requirements

For this chapter, you'll need a `conda` environment setup. If you set this up in earlier chapters, the same `conda` environment will work fine. To get into the `conda` environment, you can run the following command:

```
conda activate deepfakes
```

If you have not created a `conda` environment to run the code, it's recommended that you go to the Git repository and follow the instructions there. You can find the full repository at `https://github.com/PacktPublishing/Exploring-Deepfakes`.

Preparing to convert video

Conversion is not just a "one-and-done" script. It requires you to have turned a video into a series of frames and run C5-face_detection.py on those frames. This gets the data that you need for the conversion process in the right form. The conversion process will require the full extraction of every frame, as well as the face_alignments.json file that is generated by the extract process:

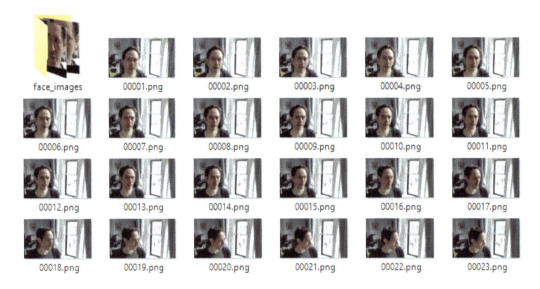

Figure 7.1 – Example of a folder that has been extracted already. Note
the face_images folder created by the extract process

If you haven't done the extract process on the video you want to convert, then you should go back to *Chapter 5, Extracting Faces*, and extract the video.

We need to do this because this is how the model knows which faces to convert. AI can detect all faces in a frame but won't know which ones should be swapped, meaning that all faces will be swapped. By running the extract process and cleaning out the faces we *don't* want to swap from the folder of extracted faces, we can control which faces get swapped.

In addition, you would probably want to include the frames you're going to convert in your training data, in what we call "fit training," which makes sure your model has some experience with the exact frames you're converting. To do this, go back to *Chapter 6, Training a Deepfake Model*, and point the "*A*" side of your model to the directory containing the frames you're going to use to convert.

> **Author's note**
>
> If you're interested in an "on the fly" conversion process that swaps all faces, you can check the exercises page at the end of this chapter, where we raise that exact question. In fact, every chapter in this section has a list of exercises for you to get your hands dirty and get the experience of writing your own code for deepfakes.

Next, let's look at the convert code.

Getting hands-on with the convert code

Like the rest of the chapters in this section, we'll be going through the code line by line to talk about how it works and what it's doing.

Initialization

Here we will initialize and prepare the code to run the convert process:

1. Like all Python code, we'll start with the imports:

    ```
    import os
    from argparse import ArgumentParser
    import json_tricks
    import torch
    import cv2
    import numpy as np
    from tqdm import tqdm

    import face_alignment
    from face_alignment.detection.sfd import FaceDetector
    from face_alignment import FaceAlignment, LandmarksType
    from lib.bisenet import BiSeNet

    from lib.models import OriginalEncoder, OriginalDecoder
    ```

    These libraries are all ones we've already seen in previous chapters. This is because the conversion process is not really doing anything too different from what we've done before. We'll see that as we go through the code to covert the face back into the original images.

2. Next, we'll check whether we're running from the command line:

```
If __name__ == "__main__":
""" Process images, replacing the face with another as
trained
      Example CLI:
      ------------
      python C7-convert.py "C:/media_files/"
  """
```

This code doesn't actually do anything but is a common way to set up a Python file to run when called. It allows you to import the script into other scripts without running those commands.

3. Next, we'll parse the arguments for the script:

```
parser = ArgumentParser()
parser.add_argument("path",
  help="folder of images to convert")
parser.add_argument("--model-path",
  default="model/",
  help="folder which has the trained model")
parser.add_argument("--cpu",
  action="store_true",
  help="Force CPU usage")
parser.add_argument("--swap",
  action="store_true",
  help="Convert to the first face instead of the
second")
parser.add_argument("--json-path",
  default="$path/face_images/face_alignments.json",
  help="path to the json data from the extract")
parser.add_argument("--export-path",
  default="$path/convert/",
  help="folder to put images (swaps $path with input
path)")
```

This code uses the standard `ArgumentParser` library from Python to parse command-line arguments. This lets us set defaults to some options and change them if we want:

```
Opt = parser.parse_args()
opt.export_path = opt.export_path.replace("$path", opt.
```

```
path)
    opt.json_path = opt.json_path.replace("$path", opt.
path)


    main(opt)
```

Here we process the arguments, add the path to applicable variables, and pass these arguments to the `main()` function, which will actually process the conversion.

We adjust the path variables to add in the default path. This lets us have JSON and export folders within subfolders of the data folder. Otherwise, each would have to be specified separately and could be in very different places. You are still able to specify a specific folder if you want, but the defaults help keep things organized.

4. We now move back up to the start of the main function:

```
def main(opt):
```

Again, this is just the main function that does the work. It doesn't actually do anything except organize our code and allow us to keep things to a normal "pythonic" operation.

5. Our next step is to make sure that the folders we're going to write into exist:

```
if not os.path.exists(opt.export_path):
        os.mkdir(opt.export_path)
```

This section checks the export path and ensures that it already exists, creating it if it doesn't.

Loading the AI

The next step is to load the AI and put it onto the device that it needs to be on:

1. First, we check whether `cuda` is available and whether the CPU override was given:

```
device = "cuda" if torch.cuda.is_available() and not opt.
cpu else "cpu"
```

This code checks whether `cuda` is enabled in PyTorch, and if it is and the user hasn't disabled it with a command line switch, and enables `cuda` for the rest of the code accordingly.

2. Next, we build the AI models with the following code:

```
encoder = OriginalEncoder()
decoder = OriginalDecoder()
```

This code establishes the encoder and decoder. Unlike when we were training, we only need one decoder. This is because training requires both faces to be able to learn successfully, but once trained, we only need the decoder for the face we're swapping in.

3. Loading model weights comes next:

```
encoder.load_state_dict(torch.load(
  os.path.join( opt.model_path, "encoder.pth")).state_
dict())

if not opt.swap:
  decoder.load_state_dict(torch.load(
    os.path.join(opt.model_path, "decoderb.pth")).state_
dict())
else:
  decoder.load_state_dict(torch.load(
    os.path.join(opt.model_path, "decodera.pth")).state_
dict())
```

This code loads the weights from the trained model. We first load the encoder weights. These are always the same, so they get pulled in from the encoder.pth file, which holds the trained weights for the encoder.

For the decoder, by default, we want to load the "b" weights, which are stored in the decoderb. pth file, but you may want the "a" face to be swapped into the "b" image, so we included a command line switch that will load the "a" weights from the decodera.pth file instead. The weights work identically and correlate to the faces that were used to train originally. The exact order doesn't matter since we included the swap flag, but only one direction could be default, so the "b" face onto the "a" image is the assumption unless overridden here.

No matter which decoder is loaded, we first load the weights and then assign them to the state dictionary of the model. PyTorch handles all the specifics of getting the dictionary loaded as matrices and into a form ready to handle tensors.

4. Next, we move the models to the GPU:

```
If device == "cuda":
  encoder = encoder.cuda()
  decoder = decoder.cuda()
```

If the device is set to cuda, we load the models onto the GPU. To do this, we tell PyTorch to use cuda on the models, which will handle the nitty-gritty of moving the models from the CPU to the GPU.

Preparing data

Next, we need to get the data loaded and into a format that PyTorch expects:

1. First, we load the alignment data from a file:

    ```
    with open(os.path.join(json_path), "r", encoding="utf-8")
    as alignment_file:
      alignment_data = json_tricks.loads( alignment_file.
    read(), encoding="utf-8")
      alignment_keys = list(alignment_data.keys())
    ```

    This code loads the alignment data from the JSON file saved by the extract process. This file includes all the information that we need to be able to pull the face from the original image, convert it, and paste it back into the image. This uses the information from the extraction instead of doing it on the fly because that information is already generated when we created the training data, and re-using that data saves a lot of time as well as enables clean up and manual editing of the data.

    You can specify the JSON file to load with the data for the image data that you're converting, but, if left blank, the default locations will be looked at which, unless you changed it during extraction, should find the file.

    We use json_tricks again because of its very powerful handling of NumPy arrays, which automatically loads the arrays back into the correct datatype and matrix shape.

    > **Tip**
    > While the inclusion or the description of tools that edit these alignments are outside the scope of this book, the Faceswap project does include advanced alignment modification tools, including an advanced "manual" tool that allows click-and-drag editing of landmarks and faces.

2. The next step is to get a list of images to convert:

    ```
    list_of_images_in_dir = [file for file in os.listdir(opt.
    path)
      if os.path.isfile(os.path.join(opt.path, file))
      and file in alignment_keys]
    ```

    This code loads all the images from the folder and then filters the images by throwing out any that don't exist in the alignment data we loaded from the JSON data file. This makes sure that we have all the information to convert the image since even if a new image were added to the folder, we would need to extract information to be able to convert the file anyway.

The conversion loop

Here, we begin the loop that will go through each individual image one at a time to convert them:

1. We're now entering the loop and loading images.

    ```
    for file in tqdm(list_of_images_in_dir):
      filename, extension = os.path.splitext(file)
      image_bgr = cv2.imread(os.path.join(opt.path, file))
      image_rgb = cv2.cvtColor(image_bgr, cv2.COLOR_BGR2RGB)
      width, height, channels = image_bgr.shape
      output_image = image_rgb
    ```

    This code loads the image and prepares it for use. First, it gets the filename and extension into variables so we can use them again later. It then loads the file in **blue, green, red** (**BGR**) color order and converts it into a **red, green, blue** (**RGB**)-ordered image as expected by our AI. Then, it saves the width, height, and color channels into variables so we can use them again later. Finally, it creates a working copy of the output image so that we can swap any faces in that image.

2. Next, we start another loop, this time for faces:

    ```
    for idx, face in enumerate(alignment_data[file]
    ['faces']):
      aligned_face = cv2.warpAffine(image_rgb, face["warp_
    matrix"][:2], (256, 256))
      aligned_face_tensor = torch.tensor(aligned_face/255,
        dtype=torch.float32).permute(2, 0, 1)
      aligned_face_tensor_small = torch.nn.functional.
    interpolate(
        aligned_face_tensor.unsqueeze(0), size=(64,64),
        mode='bilinear', align_corners=False)
    if device == "cuda":
      aligned_face_tensor_small = aligned_face_tensor_small.
    cuda()
    ```

    This loop will process each face that is found in the alignment file and swap the faces that are found in it.

    The first thing it does is pull the face from the frame using the pre-computed warp matrix that was saved in the alignment file. This matrix allows us to align the face and generate a 256x256 image of it.

    Next, we convert that face image into a tensor and move the channels into the order that PyTorch expects them. The first part of the tensor conversion is to convert from an integer range of

0–255 to a standard range of 0–1. We do this by dividing by 255. Then we use `permute` to reorder the matrix because PyTorch wants the channels to be first, while OpenCV has them last.

Next, we create a smaller 64x64 copy of the tensor, which is what we'll actually feed into the model. Since we're doing this one image at a time, we're effectively working with a batch size of 1, but we need to use `unsqueeze` on the tensor to create the batch channel of the tensor. This just adds a new dimension of size 1, which contains the image we want to convert.

Finally, if we are using `cuda`, we move the smaller aligned face tensor onto the GPU so that we can put it through the model there.

3. Then, we send the image through the AI:

```
with torch.no_grad():
  output_face_tensor = decoder( encoder(
    aligned_face_tensor_small ))
```

This code does the actual AI swapping, and it's rather astonishing how small it is.

We start this section by telling PyTorch that we want to run the AI in this section without keeping track of gradients by using `torch.no_grad()`. We can save a lot of VRAM and run the conversion faster. This isn't strictly necessary here, but it is a good habit to get into.

Next, we put the tensor containing the 64x64 face through the encoder and then the decoder to get a swapped face. The encoder's output is fed straight into the decoder because we don't need to do anything with the latent encoding.

4. Here, we apply the mask to the output:

```
output_face_tensor = torch.nn.functional.interpolate(
  output_face_tensor, size=(256,256))
mask_img = cv2.imread(os.path.join(extract_path,
  f"face_mask_{filename}_{idx}.png"), 0)
mask_tensor = torch.where(torch.tensor(mask_img) >
  200, 1, 0)
output_face_tensor = (output_face_tensor.cpu() *
  mask_tensor) + ( aligned_face_tensor.cpu() * (1 -
  mask_tensor))
```

We want to apply the mask so that we don't swap a big square box of noise around the face. To do this, we will load the mask image and use it to cut out just the face from the swap.

First, we resize the swapped face up to a 256x256 image. This is done because the mask is a 256x256 image, and applying it at a higher resolution helps to get the best detail on the edge instead of downscaling the mask to 64x64.

Next, we load the mask image. To do this, we use the aligned face filename to generate the mask image filename. We then load that as a grayscale image using OpenCV's image reader:

Figure 7.2 – An example of a mask image

That image is then converted into a tensor using a cutoff point where if a pixel of the grayscale mask image's value is higher than 200 (in a range of 0-255), then treat it as a 1; otherwise, treat it as a 0. This gives us a clean binary mask where the 1 value is a face to swap and 0 is unimportant background. We can then use that mask to paste just the swapped face back into the original image.

Finally, we apply the mask to the image. This is done by multiplying the output face by the mask and multiplying the original face by the inverse of the mask. Effectively, this combines the face from the swap result with the rest of the image being pulled from the pre-swapped aligned image.

5. Next, we'll put the face back in the original image:

```
output_face = (output_face_tensor[0].permute(1,2,0).
numpy() *
    255).astype(np.uint8)
output_image = cv2.warpAffine(output_face,
    face["warp_matrix"][:2], (height, width), output_image,
    borderMode = cv2.BORDER_TRANSPARENT,
    flags = cv2.WARP_INVERSE_MAP)
```

This code section completes the face loop. To do this, we apply the face back to the output image.

First, we convert the face tensor back into a NumPy array that OpenCV can work with. To do this, we grab the first instance in the tensor; this effectively removes the batch size dimension. Then, we'll use permute to move the channels back to the end of the matrix. We then have to multiply by 255 to get into the 0–255 range of an integer. Finally, we convert the variable into an integer, making it usable in OpenCV as a proper image.

We then use OpenCV's cv2.warpAffine with a couple of flags to copy the face back into the original image in its original orientation. The first flag we use is cv2.BORDER_TRANSPARENT, which makes it so that only the area of the smaller aligned face gets changed; the rest of the image is left as it was. Without that flag, the image would only include the replaced face square; the rest of the image would be black. The other flag we use is cv2.WARP_INVERSE_MAP, which tells cv2.warpAffine that we're copying the image back into the original image instead of copying part of the original image out.

With those two flags, the aligned image of the face gets put back into the correct place of the original full-sized image. We do this with a copy of the original image so we can copy multiple faces onto the image if multiple faces were found.

6.   Finally, we output the new image with the faces swapped:

```
output_image = cv2.cvtColor(output_image, cv2.COLOR_
RGB2BGR)
    cv2.imwrite(os.path.join(opt.export_path,
      f"{filename}.png"), output_image)
```

The last step of the image loop is to write the images in memory to separate image files. To do this, we first convert the images back into the BGR color order that OpenCV expects. Then we write the file out to the extract path using the same original filename with a PNG file type.

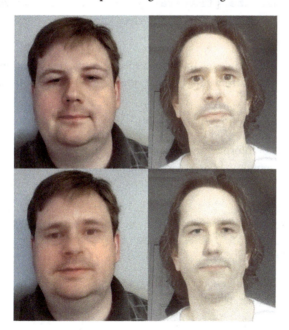

Figure 7.3 – Example of originals (top) and swaps (bottom) of
Bryan (top left) and Matt (top right), the authors

Now that we've run the conversion on the frames, we need to turn the images back into a video. Let's do that now.

Creating the video from images

The conversion code included produces swapped images, but if we want to create a video, we'll need to combine the output into a video file. There are multiple options here, depending on what you want to include:

- The following is for including just the images:

```
ffmpeg -i {path_to_convert}\%05d.png Output.mp4
```

This command line will convert all the frames into a video with some default options. The Output.mp4 file will include the frames but won't include any audio and will be at a default frame rate of 25 frames per second. This will be close enough to accurate for videos that came from film sources, such as Blu-rays or DVDs. If the video looks too fast or too slow, then your frame rate is incorrect, and you should look at the next option instead to match the correct frame rate.

- Including the images at a specific frame rate:

```
ffmpeg -framerate {framerate} -i {path_to_convert}\%05d.
png
   Output.mp4
```

This command line will include the frames at a specific frame rate. The frame rate is something you'll need to find yourself from your original video. One way to do it using ffmpeg is to run the following code:

```
ffmpeg -i {OriginalVideo.mp4}
```

This will output a lot of information, most of which will not be useful to us. What we need to do is look for a line containing the "stream" information for the video. It will look something like this:

```
Stream #0:0(eng): Video: hevc (Main) (hvc1 / 0x31637668),
yuvj420p(pc, bt470bg/bt470bg/smpte170m), 1920x1080, 13661
kb/s, SAR 1:1 DAR 16:9, 59.32 fps, 59.94 tbr, 90k tbn,
90k tbc (default)
```

The important information here is where it says 59.32 fps. In this case, we'd want to put the 59.32 into the framerate of the ffmpeg command.

This option still won't include any audio.

- Including audio with the video:

```
ffmpeg -i {path_to_convert}\%05d.png -i {OriginalVideo.
mp4}
   -map 0:v:0 -map 1:a:0 Output.mp4
```

This command will convert the video while also copying over audio from the original video file. It's important to use the exact same file for the audio to line up. If the audio doesn't line up correctly, you may want to double-check the frame rate and the number of frames.

Summary

In this chapter, we ran the convert process on a folder full of images, replacing the faces using a trained model. We also turned the images back into a video, including changes to account for frame rate and copying audio.

We started by going over how to prepare a video for conversion. The convert process requires data created by the extract process from *Chapter 5, Extracting Faces*, and a trained AI model from *Chapter 6, Training a Deepfake Model*. With all the parts from the previous chapters, we were ready to convert.

We then walked through the code for the conversion process. This involved looking at the initialization, where we covered getting the Python script ready to operate. We then loaded the AI models and got them set up to work on a GPU if we have one. Next, we got the data ready for us to convert the faces in each frame. Finally, we ran two nested loops, which processed every face in every frame, swapping them to the other face. This part gave us a folder filled with swapped faces.

After that, we looked at some commands that took the folder of swapped face images and returned them into a video form, this involved taking every frame into the video, ensuring that the frame rate was correct, and the audio was copied if desired.

In the next section, we'll start looking into the potential future of deepfakes, with the next chapter looking at applying the techniques we've learned about deepfakes to solve other problems.

Exercises

1. We use the mask to cut out the swapped face from the rest of the image but then copy it over to the aligned face. This means that the areas of the aligned image that aren't the face also get a lower resolution. One way to fix this would be to apply the mask to the original image instead of the aligned image. To do this, you'll need to call `cv2.warpAffine` separately for the mask and the aligned image, then use the mask to get just the face copied over. You may want to view the documentation for OpenCV's `warpAffine` at `https://docs.opencv.org/3.4/d4/d61/tutorial_warp_affine.html`.

    Be sure to account for the fact that OpenCV's documentation is based on the C++ implementation, and things can be a bit different in the Python library. The tutorial pages have a **Python** button that lets you switch the tutorial to using the Python libraries.

2. We rely on pre-extracted faces in order to convert. This is because a lot of the data is already processed in the extract process and is already available, allowing you to filter images/faces that you don't want to be converted. But if you're running a lot of videos or planning on running conversion on live video, it might make sense to allow conversion to run on the fly. To do this, you can combine the extract process with convert and run the extraction steps as needed before you convert. You can look at the code in `C5-extract.py` and add the appropriate parts to the convert process to enable it to work directly on the images.

3. We operated the convert process entirely on images, but it's actually possible for Python to work directly with video files. To do this, try installing and using a library such as PyAV from `https://github.com/PyAV-Org/PyAV` to read and write directly to video files instead of images. Remember that you may need to account for audio data and frame rate in the output.

4. One problem with the techniques used in this chapter is that the swapped-in face can look pretty obvious at the edges. This is because of a lack of color matching and edge blending. Both these techniques can improve the swap's edges. There are a lot of color-matching techniques available; one option is histogram matching (`https://docs.opencv.org/3.4/d4/d1b/tutorial_histogram_equalization.html`). You'll need to match the RGB channels separately. Edge blending is usually done by blurring the mask; you can accomplish this by smoothing the mask image with OpenCV (`https://docs.opencv.org/4.x/d4/d13/tutorial_py_filtering.html`). This can dull the sharp edges of the swap.

5. The results from our AI here are limited to just 64x64 pixels. There are newer models that go higher but are still limited heavily by available GPU memory and can take a lot longer to train. To get around this, you could run the output through an AI upscaler, such as ESRGAN (`https://github.com/xinntao/Real-ESRGAN`), or a face-specific restoration tool, such as GFP-GAN (`https://github.com/TencentARC/GFPGAN`). See if you can run the model's output through these before returning the face to the original image to get a higher-quality result.

Part 3:
Where to Now?

Like all inventions, the development of deepfakes is just the beginning. Now that you know how deepfakes work, where can you take that knowledge and what can you do with it? You might be surprised at how flexible the techniques can be.

In this part, we'll examine some hypothetical projects and how techniques in deepfakes could be used to make them easier, as well as solving complicated issues that might otherwise stump the average developer (you're not one of those – after all, you bought this book!). Then, we'll ask the ultimate question: what will the future bring? We'll try to answer it by looking at where generative AI might go in the near future, including looking at the limitations and challenges that these AI technologies must overcome.

This part comprises the following chapters:

- *Chapter 8, Applying the Lessons of Deepfakes*
- *Chapter 9, The Future of Generative AI*

8
Applying the Lessons of Deepfakes

The techniques in this book can be used for a lot more than face replacements. In this chapter, we'll examine just a few examples of how you can apply the lessons and tools of this book in other fields. We'll look at how to tweak and modify the techniques to use the results in new and unique ways.

In particular, we'll look at just a few techniques from earlier in this book and see how they can be used in a new way. The examples in this chapter are not exhaustive, and there are always more ways that you could implement the abilities that deepfakes bring. In this chapter, we are more focused on the technique than the specifics, but in examining the technique, we'll explore the following in new ways:

- Aligning other types of images
- The power of masking images
- Getting data under control

Technical requirements

For this chapter, there is one section with a small amount of code that demonstrates how to use a non-module Git repo for your own uses.

While this isn't part of the hands-on section of the book, we've included the code to interface with a library: PeCLR. This code is also included in the book's code repo with some additional functionality, including visualizing the points, but is just an example and is not meant to be a complete API for using PeCLR in your own project:

1. First, open Anaconda Command Prompt.
2. On Windows, hit *Start* and then type anaconda. This should bring up the following option:

Figure 8.1 – Anaconda Prompt

Click on this, and it will open an Anaconda prompt for the rest of the following commands.

3. Next, we need to clone a copy of the PeCLR library:

    ```
    git clone https://github.com/dahiyaaneesh/peclr.git
    ```

4. Download the model data.

    The library includes a copy of all the pretrained models at https://dataset.ait.ethz.ch/downloads/guSEovHBpR/. Open the link in a browser and download the files (if this URL fails, check the PeCLR library or book repository for any updated links).

    Extract the files into the data/models/ folder inside your local copy of the PeCLR repo.

5. Create a conda environment with all the PeCLR requirements installed:

    ```
    conda create -n PeCLR
    conda activate PeCLR
    pip install -r requirements.txt
    ```

    This will create an Anaconda environment with all the libraries that PeCLR needs to run.

    Additionally, this will install a Jupyter notebook. Jupyter Notebook is a useful tool for real-time coding. To run a cell, click on it and then either hit *Shift + Enter* or click on the **Play** triangle button. Jupyter will run that one chunk of code and then stop, allowing you to change the code and rerun it at will.

6. Copy the PeCLR.ipynb file into the cloned repo folder.

    If you want to follow the Jupyter Notebook file, you can just copy the file from the book's repo into the folder that you cloned PeCLR into earlier. This will save you from having to retype everything.

7. Open Jupyter Notebook and access it with a browser:

    ```
    python -m jupyter notebook
    ```

    This will run Jupyter Notebook. If you're running the command on the same computer that you're using it on, it should also automatically open your browser to the running Jupyter Notebook instance, and you'll be ready to go. If not, you can open your favorite browser and go to http://<jupyter server ip>:8888/tree to access it.

The usage of this code will be explained when we come to the *Writing our own interface* and *Using the library* parts of the next section.

Aligning other types of images

Aligning faces is a critical tool for getting deepfakes to work. Without the alignment of faces, we'd be doomed with extremely long training times and huge models to correct the faces. It's not a stretch to say that without alignment, modern deepfakes would effectively be impossible today.

Alignment saves time and compute power by removing the need for the neural network to figure out where the face is in the image and adapt for the many different locations the face may be. By aligning in advance, the AI doesn't even need to learn what a face *is* in order to do its job. This allows the AI to focus on learning the task at hand, such as generating realistic facial expressions or speech, rather than trying to locate and correct misaligned faces.

In addition to improving the efficiency of the training process, aligning faces also helps to improve the quality and consistency of the final deepfake. Without proper alignment, the generated faces may appear distorted or unnatural, which can detract from the overall realism of the deepfake.

In fact, alignment doesn't just apply to faces. You could use it for hands, people, animals, or even cars and furniture. In fact, anything that you can detect with defined parts can be aligned. For this to work, you need to somehow find points to align. For example, with hands, this could be the individual fingers.

While this works with any object, we will focus on a single example case. Here is an example process on how you could align hands. Other objects could be aligned in the same way. You'll just want to follow the same steps but replace the hands with whatever object you want to align.

Finding an aligner

First, we need to find a way to identify the points of the hand that we're interested in aligning with. For this, we need to do something called pose estimation. We could develop this ourselves using YOLO (`https://github.com/ultralytics/yolov5`) or another object detection tool that would identify some point, such as the tips of the fingers. You might have to do some heuristics to order them properly so that you can align with them.

However, better than developing this ourselves, we could use a library that does this for us. When I want to find a library or code that does a particular task, the first place I look is **Papers with Code** (`https://paperswithcode.com/`). This site has all sorts of software projects based on various AI tasks. In our example, they have a section specifically for hand pose estimation (`https://paperswithcode.com/task/hand-pose-estimation`), which lists a variety of benchmarks. These are tests that the code has been tested against. This lets you see not only the libraries that will do what you want but even show you the "best" ones.

Right now, the best result is **Virtual View Selection**, which is located at `https://github.com/iscas3dv/handpose-virtualview`. Unfortunately, this one has a restrictive "no commercial use" license. So, we'll actually skip it and go to the next one, **AWR: Adaptive Weighting Regression for 3D Hand Pose Estimation**, which can be found at `https://github.com/Elody-07/AWR-Adaptive-Weighting-Regression`. This one is MIT-licensed, which is an open license that lets you use the software even for commercial purposes, but only works on depth images. *Depth* refers to the distance between the camera and the object in the image. These images are useful for tasks such as hand detection, but unfortunately, they require special cameras or techniques to get right, so we will have to skip this one too.

Many of the others only work on depth images, too. However, if we keep looking through the posted options, we should come across **PeCLR: Self-Supervised 3D Hand Pose Estimation from monocular RGB via Equivariant Contrastive Learning**, which has an MIT license and works on standard RGB (color) photos. You can download it at `https://github.com/dahiyaaneesh/peclr`.

> **Author's note**
>
> While in this section we're just using the code and treating it like a library, in reality, the `PeCLR` code (and the other projects listed) was released as a part of an academic paper. It is not the intention of the authors to diminish that work, as academic work drives a lot of innovation in the AI field. However, this section of the book is about how to *implement* ideas, and that means using the code without necessarily paying attention to the innovations. If you're interested in a deep dive into what exactly `PeCLR` is doing, we recommend that you read the paper, which is linked in the Git repo readme.

Using the library

The `PeCLR` library has all the models and tools needed to do detection and pose estimation for the hands but not all the code to run on the external image. Unfortunately, this is very common in academic research-style projects that are often more interested in you being able to validate the results that they have already published instead of letting you run it on new data. Because of this, we'll need to actually write some code to run our images through their model.

Finding the best place to interface with the existing code can be hard if they don't provide an easy-to-use API. Since `PeCLR` was an academic project, there is no easy API, and we'll need to find our own place to call their code with our own API substitute.

Writing our own interface

The code to run the model on the validation data is only partially usable for our situation since the dataset that they were using expects data to be in a certain format, which would be hard to recreate with our data. Because of this, we'll start from scratch and call the model in our own code.

Let's get started with this:

1. First, we'll want to import all the libraries we're using:

    ```
    import torch
    import torchvision.models as models
    import cv2
    from PIL import Image
    from matplotlib import pyplot as plt
    import numpy as np
    import os
    import json
    from easydict import EasyDict as edict


    from src.models.rn_25D_wMLPref import RN_25D_wMLPref
    ```

    The preceding code imports all the libraries we're going to need. Most of these are standard libraries that we've used before, but the last one is the model that PeCLR uses to do the actual detection. We've imported that one, so we can call it with the image to run the model.

2. Next, we'll load the model from PeCLR:

    ```
    model_path = 'data/models/rn50_peclr_yt3d-fh_pt_fh_
    ft.pth'
    model_type = "rn50"
    model = RN_25D_wMLPref(backend_model=model_type)
    checkpoint = torch.load(model_path)
    model.load_state_dict(checkpoint["state_dict"])
    model.eval().cuda()
    ```

    The preceding code loads the model from the model data that PeCLR provides. To do this, first, we define the model path and type. Then, we pass the model type to generate an appropriate model. Next, we load the checkpoints and copy the weights into the model. Finally, we prepare the model for evaluation and set it to run on the GPU.

3. Next, we'll load the image and prepare it for the model:

    ```
    image=io.imread(
        "https://source.unsplash.com/QyCH5jwrD_A")
    img = image.astype(np.float32) / 255
    image_mean = np.array([0.485, 0.456, 0.406])
    image_std = np.array([0.229, 0.224, 0.225])
    ```

```
img = np.divide((img - image_mean), image_std)
img = cv2.resize(img, (224,224))
img = torch.from_numpy(img.transpose(2, 0, 1))
img = img.unsqueeze(0).float().cuda()
```

This code prepares the image. It does this by, first, loading it with the `SciKit` image loader, which, unlike `OpenCV`, can directly handle URLs or local files. It then calculates an adjustment for restoring the model's coordinates to the ones that match the full image size. It does this by dividing 224 by the height and width of the image. Then, we convert the image data into a floating point with a range of 0–1. We then normalize the images by dividing them by a standard deviation and subtracting a mean. This brings the images down to a range that the model expects. Then, we resize the image to 224 x 224, which is the image size that the model expects. We then convert the image into a tensor and get it in the order Pytorch uses, with the channels first. Finally, we add another dimension at the front to hold the batch and convert it into a 32-bit floating point on the GPU.

This all prepares the image for the model to be run on it.

4. Next, we run the model on the image and get 2D coordinates out of it:

```
with torch.no_grad():
    output = model(img, None)
kp2d = output["kp25d"][:, :21, :2][0]
height, width = image.shape[:2]
kp2d[:,0] *= width / 224
kp2d[:,1] *= height / 224
```

This code first runs the image through the model without generating training gradients. To do this, we pass the image and the **camera intrinsics** (see the *Author's note*). In our case, we don't know the camera and also don't really care, so we just pass an empty `None` value, which will use a default camera intrinsics matrix.

> **Author's note**
>
> *Camera intrinsics* is a fancy term, but it just means the details of your camera. In the case of PeCLR, it wants a matrix that details how large the pixel space is so that it can attempt to guess the depth information from the 2D image. We don't need the depth information, so we can let PeCLR create a default matrix instead of giving it one.

Next, the code takes just the 2D alignment points. We don't need 3D points since we're aligning in 2D space. If we were working with a depth image, we may have wanted the third dimension, but we aren't and we don't need that for our scenario.

Next, since the model was given a small 224 x 224 image, we're going to adjust those coordinates to match the width and height of the original image. To do this, we divide the coordinates by 224 and multiply the result by the original image sizes.

Using the landmarks to align

In this case, the library will mark the joints and tips of every finger and the thumb and one point near the "middle" of the hand. Unfortunately, the point in the middle of the hand is not well defined and could be anywhere from the actual middle to the wrist, so we wouldn't want to use it for alignment. The joint and fingertip locations are going to be more consistent, so we can use those for the alignment process:

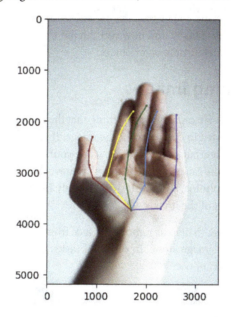

Figure 8.2 – A hand detected and marked with detection from PeCLR
(original photo by Kira auf der Heide via Unsplash)

> **Tip**
>
> Aligners are never perfectly accurate. Some can vary by a significant amount, but it's not a matter of perfect results every time. Any alignment, even an imperfect one, has benefits for neural network training as it normalizes the data into a more reliable format. Any work done before the data is given to the AI means that is one less task the AI has to waste effort on doing.

Once we can get some known points on an image, we can scale, rotate, and crop the image so that it's in the orientation that you want and run it through the detector to get a list of points. If you can, I recommend running several images through and averaging the points together. That way, you can reduce any variations in the hand images and get better alignments.

Once you have the points that you want to align to, you can use them to generate aligned images using the **Umeyama** algorithm. The algorithm just needs two sets of points, a known "aligned" set and a second set that you can convert into an aligned set. Umeyama returns a matrix that you can feed into an Affine Warp function to get a final aligned image. See *Chapter 5*, *Extracting Faces*, for a hands-on code example of how to do this.

Once we have the aligned hand images, you can do whatever it was you were planning on doing with them, be that displaying them or using them for your AI task. It's even possible that, once you have your aligned data, you can use it to train a pose detection model of your own to get even better alignment results. This process of using AI-processed data to feed into a model to make that model better is called **bootstrapping** and, with proper supervision, is an invaluable technique.

Alignments are a critical part of deepfakes, and now you can use them in other fields. Next, we'll look at how to use masking to get clean cut-outs of an object.

The power of masking images

When you take a photograph, you are capturing everything that the camera sees. However, the chances are that you're not equally interested in every part of the image. If you're on vacation, you might take a selfie of yourself in front of a waterfall, and while you value yourself and the waterfall, you care less about the cars or other people in the image. While you can't remove the cars without adding something into the gaps for your vacation photos, sometimes, you're only interested in the main subject and might want to cut it from the rest of the image.

With deepfakes, we can use a mask to help us remove the face from the image so that we replace only the face and leave the rest of the image alone. In other AI tasks, you might have similar needs but different objects that you want to cut out:

Figure 8.3 – An example of the mask used in the deepfake process

Next, let's look at other types of masking.

Types of masking

Masking out an image can be useful in a lot of tasks. We only used it as a part of the conversion process in a step called **composition**. This is only part of the power of masking. It can be used to guide **inpainting**, which is the process by which you fill in gaps in an image to erase an object. You could also do it to an image before it gets fed into an AI to make sure that the AI can focus on the important parts.

In order to mask an image, you need to get some sort of idea of what part of an image you want to be masked. This is called **segmentation**, and it has a lot of sub-domains. If you want to segment based on the type of object, it would be called **semantic segmentation**. If you wanted to segment the image based on the subject, it is called **instance segmentation**. You can even use **depth segmentation** if you have a solid depth map of the image. Unfortunately, deciding which type of segmentation you need requires special attention in order to find the right tool.

Finding a usable mask for your object

Libraries such as **PaddleSeg** (`https://github.com/PaddlePaddle/PaddleSeg`) have special tools that let you do multiple types of segmentation. They even have an interactive segmentation system that lets you "mark" what you want to segment like Photoshop's magic wand tool. Following that, you might need to use that data to train a segmentation model that is capable of masking that particular type of object in new contexts.

To find the best method of masking the given object you're interested in, you should probably start with a search for the item that you want to mask and segment. For some objects such as faces, cars, people, and more, there are prebuilt models to segment those objects.

But if a segmentation model doesn't exist for the particular object you're interested in, there's no need to despair. Newer models such as **CLIP** (`https://github.com/openai/CLIP`) have opened up whole new opportunities. CLIP is made up of a pair of AI models that connect language and images together. Because of the shared nature of CLIP, it's possible to learn the difference between objects based on their text descriptions. This means that libraries such as **CLIPseg** (`https://github.com/timojl/clipseg`) can use language prompts to segment objects in an image.

Examining an example

Let's look at an example. Let's say that you wanted to count the cars parked in a parking lot and see whether any spaces are still available, but all you have is a webcam image of the parking lot from above. To do this, you need to know which parts of the image are cars and which are empty parking spots. This task mixes both semantic segmentation and instance segmentation but uses them together.

The first step would be to mark out each parking spot in the image to define which spots you want to look at. You could pick a single spot in each parking spot or define them by the whole area. Either way, you'll probably want to do this manually since it is unlikely to change often enough to justify having the computer do it.

Now that you know where the parking spots are in the image, you can start looking for cars. To do this, we'd want to look around for a good neural network trained to do this task. In our case, we can check this example from Kaggle user Tanishq Gautam: `https://www.kaggle.com/code/ligtfeather/semantic-segmentation-is-easy-with-pytorch`. This page gives pretrained models and a solid guide for how to segment cars.

While this model does not do instance segmentation (giving each car a different "color" or tag), we can use it to count the cars anyway. We can do this because we've already established that we're counting cars in a parking lot and that we have manually marked each parking spot.

Then, we can simply use the segmentation model to detect any cars and see whether they overlap with the parking spots. If we only marked a single point for each parking spot, we could simply check each point and see whether it is segmented as a "car" or as "background" (a common term used for anything that we're not segmenting).

If we marked the whole area of the parking spot, then we might want to calculate a minimum coverage for the spot to be considered "taken." For example, we want to ensure that a motorcycle in a parking spot gets counted, but a car that is slightly over the line shouldn't be counted. You can set a minimum threshold of the parking spot's area, and if it's filled beyond that with a "car," then you mark the whole spot as taken.

Additional functionality is possible too. For example, you could even detect whether a car is in a prohibited area by checking whether an area segmented as a "car" is not inside a parking spot. This could be used to automatically alert a parking enforcement officer to check the lot.

Now that we've gotten a good hold on our masking, let's look at how we can manage our data.

Getting data under control

There's a common saying in the AI community that an ML scientist's job is only 10% ML and 90% data management. This, like many such sayings, is not far from the truth. While every ML task is focused on the actual training of the model, first, you must get your data into a manageable form before you can start the training. Hours of training can be completely wasted if your data isn't properly prepared.

Before you can start training a model, you have to decide what data it is that you're going to train it with. That data must be gathered, cleaned, converted into the right format, and generally made ready to train. Often, this involves a lot of manual processes and verification.

Defining your rules

The most important thing in the manual process is to make sure that all your data meets your requirements and meets a consistent level of quality. To do this, you need to define exactly what "good" data means. Whether you're annotating your data or gathering large amounts, you should have a standard way of doing whatever it is you're doing.

For example, let's say you're annotating a dog versus cat dataset and you want to put all the pictures into one of two buckets, one bucket consisting of all the dog images and the other bucket consisting of all the cat images. What happens when an image contains both a cat and a dog? Do you put it in the bucket that is more prominent? Do you exclude it from your dataset? Do you edit the image to remove one of the animals? Do you crop it into two images so both animals end up in the dataset?

It's important to have a consistent set of rules for these situations. That ensures that your data is appropriate for the purposes. You don't want to have these edge cases happen in different ways each time, or it will upset your training and raise confusion when you're trying to fix issues.

Evolving your rules

Also, it's important to have a plan for what happens when you change your rules. As you go forward with your data management, it's almost inevitable that you'll find that you want to make some tweaks or changes to the rules based on what data you find, how well your training process goes, and whether your final use case changes.

Looking back at our example, let's consider the case where you decided to exclude any images that contained both cats and dogs but other animals were fine as long as the image also contained a cat or dog. What happens when you decide that you want to add rabbits to your cat/dog detector? This means not just that you add a new bucket for rabbit images but also that you have to re-process all the existing images that you have gone through to make sure that any cat or dog images that also contain rabbits get removed. What about if you find out that guinea pigs in your cat and dog buckets are being flagged as rabbits? These processes need to be considered as you manage your data.

Dealing with errors

Errors happen, and small amounts of bad data getting through into your dataset is inevitable. However, there is a fine line between a few harmless errors and a large enough error to completely invalidate the training. Because of this, it's often best to get a second (or third) set of eyes on the data as a part of the standard process. Sometimes, this isn't possible, but in any situation where it is possible, it's invaluable. A second set of eyes could find flaws in your data or even your methodology. What if you were tagging a particularly weird-looking animal as fine in your dog data but a second person identified it as a rare breed of cat?

I also recommend automating as much of your data gathering as possible. When you can cut a human out of the loop, you'll not only save timebut you'll also prevent errors and mistakes. Time spent automating a data process will almost always pay back dividends in time down the line. Even if you think that there is no way that a process will take less time to automate than to do it manually, you should consider the errors that you will avoid. Well-written automation code can be reused later for future projects, too. Anything that can be reused should be automated so that the next time you need to get it done, there is a tool to handle it for you.

Summary

In this chapter, we looked at how you can apply the lessons and techniques of deepfakes to other environments. First, we examined how to align other types of images, using hands as an example. Then, we looked at the different types of masks and considered using them in a parking lot monitoring solution. Following this, we examined data management and considered how a dataset to detect different animals might be built.

This process of figuring out how to apply techniques in new environments used throughout this chapter is itself a valuable technique that can help you throughout your development career, especially if you're going to work at the edge of your computer's capabilities like AI does now. Sometimes, the only difference between a successful project and an impossible one is the technique you borrow from a previous project.

In the next chapter, we're going to look at the potential and future of deepfakes and other generative AIs.

9
The Future of Generative AI

While it sometimes might feel like we're already living in the future with deepfakes and AI-generated images, the technology behind them is really just beginning to take off. As we move forward, the capabilities of these generative AIs will only become more powerful.

This chapter is not unbounded futurism but will instead look at specific generative AIs and where they are improving. We will examine the following technologies and how they are changing. We'll discuss the future of the following areas of AI:

- Generating text
- Improving image quality
- Text-guided image generation
- Generating sound
- Deepfakes

Generating text

Recently, text generation models made a major impact when they came into the public consciousness with OpenAI's success with ChatGPT in 2022. However, text generation was among the first uses of AI. Eliza was the first **chatbot** ever developed, back in 1966, before all but the most technically inclined people had even seen a computer themselves. The personal computer wouldn't even be invented for another 5 years, in 1971. However, it's only recently that truly impressive chatbots have been developed.

Recent developments

A type of model called **transformers** is responsible for the recent burst in language models. Transformers are neural networks that are comprised entirely of a layer called an **attention layer**. Attention layers work sort of like a spotlight, focusing on the part of the data that is most likely to be important. This lets transformers (and other models that use attention layers) be a lot deeper without losing "focus" (though that's not actually the technical term for it, it's a good metaphor for the effect).

Thanks to the ability to make very deep models, transformers excel at tasks that need to pull meaning from complicated structures. This is a perfect match for language tasks and other long sequences where a word's meaning depends heavily on those words around it (think about the difference between "*I ate an orange fruit*" and "*I ate an orange*"). Transformers are used heavily for tasks such as translation, but they also are excellent for other tasks such as answering questions, summarizing information, and the subject of this section: text generation.

Most language tasks require not just the ability to "understand" the text but also to create new text. For example, in a summarizing task, you cannot simply take the important words from a document and copy them out to the user. The words would be meaningless, as they'd be in an arbitrary order and have none of the important context that language is built on. So, these models must also create realistic and understandable output in the form of properly constructed sentences for the user to interpret.

> **Author's note**
>
> There is a lot of debate as to whether you can use a uniquely human concept of understanding to describe anything a computer does. One famous example is the so-called **Chinese Room** thought experiment by John Searle.
>
> The basics of the thought experiment are this: imagine a room where there is a huge array of filing cabinets. In each of these filing cabinets, there are cards that pair a certain input to a certain output. The room can take as input and output other cards on which Chinese characters are written. Inside the room, there is someone (or something) who knows no Chinese, but by matching the incoming cards with the cards in the filing cabinets, then copying the card's output characters onto another card, can output perfect Chinese.
>
> In that context, can you say that the machine (or the person or thing inside the machine) "understands" Chinese? What about the room itself? The question has led to a lively debate among philosophers, computer scientists, linguists, and more.

Building sentences

Eliza, that first chatbot, built its sentences by simply repeating back words from the input sentences into one of a number of pre-built sentences that were written carefully to let those words be put into them. This process of parroting parts of the user's responses back was good enough to serve as a very simple Freudian counselor, which many people felt helped them to get through some situations.

Today's generative models instead build sentences with extremely advanced sentence structures. They do this by building sentences in a way not unlike a jigsaw puzzle. The model uses a concept called a **beam** to search for the word that fits best into a given part of a sentence, much like how a jigsaw solver searches over a set of pieces to find the ones that best fit the available space. Unfortunately, this is where the metaphor becomes strained. A single beam keeps track of just one sequence, and unlike a normal jigsaw, a sentence can be made of a near-infinite combination of words. Because of this, a model needs to keep track of several beams as the best scoring beam may change as the sentence gets completed. More beams lead to a better-generated sentence, both **grammatically** and **semantically** (that is, both the structure and meaning) but require more computation as the model solves a larger number of sentences at the same time.

The future of text generation

As we move into the future, these models will improve in two ways; computational availability will increase, making it feasible to get larger models with larger beam searches, and design efficiency will allow for better-designed models that can do more with the same resources. You can think of this as something like packing boxes into a truck: you can only fit so many boxes in a truck until you need to either get a bigger truck or find a way to make the boxes smaller. Transformers are an example of design efficiency, while the fact that computers always get faster and better is an example of computational availability.

It may seem that text generation has endless grounds for improvement, but there are questions as to how sustainable the long-term growth of language models may be. There are concerns that we might reach the limits of **large language models (LLMs)** relatively quickly. Multiple constraints exist that may limit the models.

The first limit on language models is that the largest models already require weeks or months to train, spread across thousands of powerful computers. The biggest limitations of these types of models may be economics. There simply might not be enough demand to keep growing the models as the costs continue to skyrocket.

The second major limit to language models is far harder to get past: there simply might not be enough data to keep growing. LLMs require a massive amount of data to be able to learn languages, and we are quickly approaching the point where they could be trained on the entire output of humanity's written history and still not have enough data to fully train the model. Deepmind published (`https://arxiv.org/abs/2203.15556`) a study of LLMs that found that the compute and data needed to completely train a model must scale up in tandem. That means that if you double the amount of compute a model needs to train, you should also double the data. This means that it's entirely possible that models will soon be limited in how well they can learn language because they've been trained on everything ever written by humanity.

Both of these problems have potential solutions, but they're nowhere near as easy to come by as throwing more compute or smarter model designs at the problem. You might ask something such as *"but how can humans learn language without reading every text humanity has ever written?"* or something similar. The reason is that humans are fully "intelligent," a term whose definition seems to change every time something that is not human gets close to it. In the end, it's possible that we will be unable to get a computer to fully understand language until we can get them to "understand" in a way that philosophers cannot debate. Until then, language models need a lot more data than humans to reach high levels of language use.

Text generation may have been one of the first uses of AI, but arguably a more significant use until recently has been improving image quality.

Improving image quality

The earliest known image that was taken through mechanical means is the *Niépce héliographie*, which was taken by Joseph Nicéphore Niépce in 1827. It was taken through the window of his workshop by exposing a pewter plate covered in a thin layer of a concoction made from lavender and bitumen. This plate was exposed to sunlight for several days to create a blurry, monochrome image.

Figure 9.1 – The Niépce Heliograph taken by Joseph Nicéphore Niépce in 1827

Since then, images have gotten better at capturing reality, but the process has not been *perfected*. There are always various limitations that mean that the images aren't quite accurate in color, cannot capture all the details, or cause distortions in the image. A truly perfect image is mindboggling to even consider: from a single perfect image, you'd be able to zoom into any atom even on the other side of the universe and our images simply cannot accomplish that level of detail.

Various tactics

Sometimes, the limitations of the image are more than we're willing to accept. In this case, we can try to "improve" the image by modifying it to look more like what we want. Some of these edits are relatively easy. Perhaps your photo was taken at an angle, and everything is crooked. In the age of physical photo prints, this could be fixed with nothing but a pair of scissors (and maybe a mat to hide the smaller size). Many cameras with a flash attached have built-in "red eye" removal, which masks the reflection of the flash in people's eyes by firing the flash twice in quick succession, once to make your eyes adjust to the flash and the second to actually take the photo.

Now you can fix both of these problems in software, and you can fix a lot more than crooked photos or a bright light reflecting off the back of your eye. Modern editing software can even "fix" things that people would argue aren't actually broken. There is an ongoing discussion in society whether the images on the covers of magazines should be "brushed up" to make the subject more attractive or whether the process creates more harm than good by perpetuating an unrealistic body image.

As we've reached the age of neural networks, the automatic improvement of photos has become easier than ever. There are numerous services and tools offering to improve photos, whether they have been newly taken or are older. They work on a variety of techniques but fall into several different categories, including **upscaling**, **color correction** (or **grading**), and **denoising**. Let's look at each of these briefly.

Upscaling

Upscaling involves taking an image and scaling it up. That is, making an image have a higher resolution. Modern upscaling AIs have the ability to not just increase the resolution but also the fidelity – they can make images both bigger and better. This innovation comes from learning from lots of other images about what objects, textures, and images look like both large and shrunk down. The model, after being fed an artificially downscaled image, is asked to recreate the original image and is then reviewed on how well it did.

Making an image larger and higher in fidelity can be a balancing act. It's possible to make images *too* sharp or fill in data that shouldn't be there. For example, you don't want to add wrinkles to a child, but a photo of your grandmother without her wrinkles just wouldn't look right. For this reason, most upscaling has some sort of slider or other control to set how much additional fidelity you want the AI to give your images, letting you choose case by case what level of detail should be added.

There is one type of upscaling that sidesteps this limitation, and that is **domain-specific upscaling**. A domain-specific upscaler is one that upscales only one type of content. A recently popular example of this is face upscaling. Humans are hardwired to see faces, and we will notice any issues with them very easily. But when you take an upscaler and train it on nothing but faces, it's able to learn specifically about faces and get far higher fidelity results when doing that one task instead of having to worry about upscaling everything. This lets the AI learn that "old people" tend to have wrinkles, something that a general purpose upscaler just doesn't have the space to learn.

The techniques of domain-specific upscaling require additional processes such as detection and compositing to detect the objects that the upscaler can improve and ensure that they get put back into the image correctly. This does lead to a different potential problem where the objects are of a higher quality than the rest of the image, but unless the difference is drastic, people don't seem to notice very much.

Color correction

Color correction is a process that takes place in every camera device anywhere. Taking an image from the camera sensor and turning it into an image that we can look at requires many different processes, one of which is color correction – if for nothing else than to correct for the color of the lighting that the picture was taken in. Our eyes and brains automatically correct for light color, but cameras need to do it explicitly. Even once a photo is taken, that doesn't mean that color correction is done. Most professional photographers take pictures in a "raw" format that allows them to set the colors exactly, without the loss of converting them multiple times.

> **Author's note**
>
> One lie that our cameras tell us is that an image is made up of many pixels that have red, blue, and green elements. In reality, our cameras take separate images of each color and then join them together in a process called **debayering**. In fact, most cameras cheat even more by having twice as many green sub-pixels as red or blue, as human eyes are more sensitive to details in green. The problem is that these pixels never line up perfectly when re-aligned, and poorly done debayering means that your images will have weird color artifacts at the edges of the colors.

However, color is not just an objective thing. Think about the film *The Matrix*. Have you ever noticed how green everything is inside the Matrix? That was done on purpose. This use of color is called "grading" (though confusingly, sometimes, correction is also called grading). Humans are weirdly connected to colors on an emotional level, and changing them can tweak our emotional connection to an image. For example, think about how we say "blue" to mean "sad" or how an image full of browns and reds might make you feel a chill like autumn has just begun.

Both of these color tasks can be performed by AI. There are AI tools out there that can convert an image to match a particular emotional style or to match another image that you have. Here, the trick for the AI is not in changing the colors of an image, it's in changing them in a particular way that will be pleasing to the humans involved. You probably don't want to turn your green shoes into red just because you ask the AI to turn the photo to fall colors, but the green leaves need to be adjusted.

Denoising

Noise is another of the inevitable issues with pictures that today's cameras do their best to hide, but it's a part of any image you take – especially in dark conditions. Noise reduction in cameras usually works by applying a slight blur to all pixels in the image (often before the color correction happens). This can help to smooth out any noise coming from the sensor, but it can't do a perfect job. Fancier techniques have become extremely common since modern cameras now have substantial computer power in them (and many are in our phones). In fact, most of today's phone processors have some sort of AI acceleration just so that they can improve the photos being taken.

Denoising can be done in many different ways, but a recent innovation is diffusion models. Diffusion models are really just using a **diffusion** process to train a neural network – one in which the image is given to the model with blur or noise added and it's tasked with providing the original image. This process can be made deeper, by repeating the diffusion process multiple times and asking the model to clean it up each time. This allows a model to create incredibly detailed images from very little information. Unfortunately, information loss is information loss, and while diffusion models can create new information, restoring the original information is impossible without some kind of guidance. This type of AI denoising is actually responsible for one of the innovations we'll talk about more in the *Text-guided image generation* section.

The future of image quality upgrading

There is a fundamental question that limits what an AI can do for image quality. Is it okay for the AI to "invent" a new detail that doesn't belong there? If the answer is yes, then there is no practical limit to what AI upscaling can accomplish. It's possible that we could have AI upscalers that generate the very atoms of a person's skin if we zoom in far enough. Unfortunately, the answer is generally considered to be "no." We are picky about our images and want them to be not just detailed, but accurate. This means that quality and detail will always be limited, and there isn't a whole lot we can do to do to avoid that.

That's not to say that it's impossible to improve further. It's possible that we might be able to make AIs capable of getting the details they need from other sources. For example, right now, we have video upscalers that can upscale a frame of video by borrowing data from the frames that surround them. This lets the AI look beyond the current frame to find patterns and details that it can copy back to the working frame. You might be able to upscale your old family home videos by providing photographs to fill in the missing quality. Additionally, you might have an AI that learns what a given person looked like at various points in their life (for example, your school yearbooks and family photos) that can then interpolate that data to fill in specific times of your video.

The quest for higher quality will never end, and people will always want to see the next improvement. That said, there are practical limits that prevent a given image from being improved forever. However, what if you don't care about accuracy, and you just want the greatest quality image possible? Maybe an image that is created from just a sentence or two? Well, that's what we'll look at next.

Text-guided image generation

Text-guided image generation is an interesting category of generative AI. OpenAI had several developers release a paper called *Learning Transferable Visual Models From Natural Language Supervision* (https://arxiv.org/abs/2103.00020). Though I prefer the summary title they posted on their blog *CLIP: Connecting Text and Images*. CLIP was mentioned in *Chapter 8, Applying the Lessons of Deepfakes*, but we'll talk about it some more here.

CLIP

CLIP is actually a pair of neural network encoders. One is trained on images while the other is trained on text. So far, this isn't very unusual. The real trick comes from how the two are linked. Essentially, both encoders are passed data from the same image; the image encoder gets the image, the text encoder gets the image's description, and then the encoding they generate is compared to each other. This training methodology effectively trains two separate models to create the same output given related inputs.

That might still sound confusing, so let's look at it another way. Imagine a room where there are hundreds of little boxes in a grid. Two men take turns in the room; one is given an image, and the other one is given a sentence. They are tasked with placing their objects in one of the boxes but with no information as to what is "correct". They can study their own objects but then must pick one of the boxes. Then, they're scored based on how close they were to each other's placement and only on that similarity, with no assumptions that there was a better choice than the ones the two men picked.

The idea is that, eventually, they'll settle on a set of rules where both the image and the text description of the object will be placed in the same boxes. This metaphor is very close to how it actually works, except instead of just putting the object in one box, they score the text and image on how they fit into each of the hundreds of boxes.

When fully trained, similar images should score similarly, similar text descriptions should be scored similarly, and matching image and text descriptions should be scored similarly:

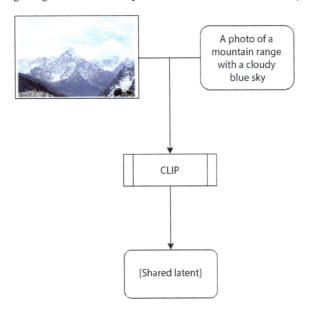

Figure 9.2 – An example of CLIP's shared latent space (Mountain photo by Rohit Tandon via Unsplash)

Image generation with CLIP

So, now we have a model that turns an image and its associated text into a **shared latent space**. How does this generate images? Well, that's where other tricks come in. We're going to gloss over the specifics here because there are so many competing models that have come out recently that all do things a little differently and writing about each one in depth would be a book of its own, but the basic idea is similar between them, so let's go with that.

Okay, so to train the model, the first step is to get the CLIP latent. Some models use both the image and the text embeddings, some go for one or the other at random, while others pick just the text embedding. No matter which latent they use, the process is similar. After generating the CLIP embedding, a model is trained to turn that embedding into an image.

Here is a quick summary of just a few of the big text-to-image models on the market right now:

- Stability AI's **Stable Diffusion** uses a repeating diffusion process that starts with a random noise pattern (or, in image-to-image mode, an image) and then iterates the diffusion process over and over in the latent space and uses a final decoder to convert the embedding back into an image

- Google's **Imagen** works with a similar method, but the diffusion process occurs in the image's pixel space, and there is no need for a final conversion back into an image space

- OpenAI's **Dall-E** is, instead, a transformer-based model that generates an image by passing the text representation through a number of attention layers and other layers to create the image output

The effective result in each case is an image that should match the text given to guide the generation.

So, where will image generation go from here?

The future of image generation

Many other techniques have been developed and announced, and just keeping up with the various releases is a significant endeavor beyond the scope of this book. The really incredible thing about all these different techniques is that they were all developed in rapid succession. The pace seems to be continually accelerating and doubtless shortly after this book comes out, all of this information will be superseded by a new innovation.

This field is rapidly developing, so it's impossible to predict where future limitations might show up. It's possible that we're near the tail end of this current burst of development, and we'll see a quite few years before a new innovation. But it's also possible that we'll see a rapid development of new functionality that makes today's image generators look primitive within a few months.

The limitations that abound right now are not about resolution, detail, or even quality: it's simply a matter of learning the best way to get results. Thanks to the public distribution of Stable Diffusion, there is a huge community of people experimenting with the technology, and it's nearly impossible to go more than a couple of days without some new technique or idea coming forward that promises new frontiers. Things such as **Textual Inversion**, **LoRA**, and **DreamBooth** offer new ways to tune Stable Diffusion results on consumer hardware, while new functionality such as **inpainting** and **outpainting** allow you to fill in or expand an image, respectively.

As time goes on, we'll get more comfortable with using these image-generation models, and we'll be better able to design them to maximize their performance. At that point, we'll be able to predict what limits image generation has. At this point, we're just too close to the mountain directly in front of us to see the top.

Generating sound

Sound generation is another one of those fields we could keep subdividing down and down until all the room we have in the book is taken up by headings listing the different methods of sound generation. For the sake of brevity, we'll group them all here and cover a few big subfields instead.

Voice swapping

The first thing that most people think about when they learn they can swap faces is the question of whether they can swap voices too. The answer is quite unsatisfying: yes, but you probably don't want to. There are AIs out there that can swap voices but they all suffer from various problems: from sounding like a robot, to lacking inflection, to not matching the person involved, to being very expensive and exclusive. If you're doing anything with even moderate production value, you'll get much better use out of *natural* intelligence: finding an impressionist who can do an impersonation of the voice. AI technology is just not there (yet). Even *Fall*, a movie that famously did "voice deepfakes" actually just re-recorded the voices in the studio (a process called automated dialog replacement or looping), and then the AI was used to adjust their mouths to match the new dialog.

Voice-swapping technology will undoubtedly improve, and there are companies that offer very high-quality voice models for various projects, but they're very expensive and still a niche use case. However, if the demand remains, it's inevitable that someone will come out with a new technique or improvement that will make high-quality voice swapping as easy as other deepfakes.

Text-guided music generation

This technique works a lot like text-to-image generation. In fact, most of the implementations actually use a text-to-image generator under the hood. The basic technique is to turn audio into a spectrogram image of the audio and then train a text-to-image model on it. This means that the same image generators can be used to generate audio spectrograms that can then be converted back into an audio stream.

This technique is lossy and limits the quality of the output because the spectrogram image can only hold so much frequency. In addition, there is the problem that spectrogram images represent just a small slice of time. This means that generating the audio has to happen in small chunks a few seconds long. Riffusion (an audio generation model built on Stable Diffusion) was trained on spectrograms about 8 seconds long. This means that even if you come up with tricks such as joining multiple spectrograms, you're going to have a new generation every 8 seconds or so. It's possible to increase both the time slice and frequency by increasing the resolution of the image underlying the generation, but the limits would still be there, just higher. It's unlikely that that current technology will be able to generate full length songs of 3-4 minutes even with reasonable advances in technique.

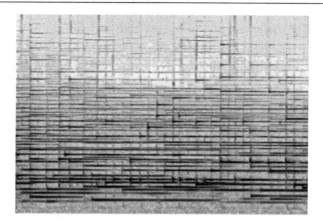

Figure 9.3 – An example of a spectrogram made by Riffusion

Music that has low-frequency resolution and changes significantly every few seconds is a far cry from what we, typically, expect as music listeners. It's likely that someone will create a text-to-music generator that will not involve the audio-image conversion process, but that would be a new creation that will probably have its own quirks and is getting into the speculative futurism that we want to avoid in this chapter.

The future of sound generation

It's all but inevitable that someone will create a tool or technique that will blow open the field of sound generation in the same ways that convolutional layers and diffusion models have done for images and transformers have done for language. When that happens, we'll see a gold rush-style burst of innovation as people adopt and extend the new technology with the same improvements that we've seen in the image and text generation fields. Unfortunately, for now, we cannot rely on the great innovations of the future, so we must ground ourselves in the now or the near future.

Modern audio techniques rely on autoencoders or spectrograms to bring the audio into a form that is usable by the AI generation techniques that we have access to now. To this end, it's possible that these methods could be improved to add more detail or information. For example, usually, spectrograms are done in black and white. What if color were added in a way that more than just linearly increased the amount that could be generated? For example, what if red were used for a repeating melody, blue for another motif, and green to show how the two interact at a much larger timescale? Could this, when paired with larger spectrograms, bring modern sound generation to the point where an entire high-quality song could be generated at once?

What if we spent the time to develop a coding system for a spoken voice that turned it into a written form that could be learned and trained into a text-generation model? If we could define an accent as some combination of how the written form gets turned into audio, could we make an AI that could copy accents like they can painting styles? There are a lot of ways that we might be able to modify audio into a form that our current AI models excel at that would enable improvements of current sound-generation AI across the board.

Not even counting the revolutionary improvements from a whole new paradigm type akin to transformers or convolutions, I think that there are years of substantial evolution available in audio if computer scientists were to put as much attention into audio as they have images and video recently.

Deepfakes

Of course, this whole book has been about the past and present of deepfakes, so it makes sense to circle back to their future at the end. There is a lot we can learn about the future of deepfakes from the other AI mentioned in this chapter. This is because, sneakily, all the parts of this chapter have been building up to this section. Deepfakes are, after all, an image generation AI that works on domain-specific images with a shared embedding.

Every area that we've explored in this chapter can be used to improve deepfakes, so let's approach them one at a time.

Sound generation

This one is quite simple and obvious. The next step after swapping a face would be to swap the voice too. If we could get a solid voice swap, then deepfakes would be taken to a whole new capability. Making music or other effects could also be useful if you were making a movie without any other people helping, but their utility would be otherwise limited (in deepfakes, other industries would doubtlessly have more use for them).

Text-guided image generation

Right now, you could consider deepfakes to be "face-guided" and perhaps that is the best solution, as it lets an actor's performance shine through, even if they're wearing a different actor's face. But text-guided postprocessing could definitely fill a need. There is a famous scene in the film *Blade Runner* where the detective, Deckard, examines an image using voice commands to explore the environment (before doing the impossible and shifting the view to the side, which could only happen by inventing new data and wouldn't be usable for finding clues). This could be seen as a prototype workflow for image modification with text-guided image generation.

Instead of generating a whole new image, we could modify parts of it using masks or filters and combine the AI changes back with the original image. This lets us do things such as write (or say) "*make him smile more*" to generate modifications to our image without replacing the whole image. This leverages the power of text guidance for some things, while still letting the face-guided swap happen.

Improving image quality

It is easy to see how improving image quality could improve deepfakes: you could improve your input data or output results to be more accurate, of higher quality, or for color correction. Image quality is critical in creating a successful deepfake, but some sources simply don't provide the data that you really need to get a quality deepfake. For example, Albert Einstein, unfortunately, passed away before HD video cameras could capture all the data that we want for a standard deepfake. Photos and some videos do exist that we could use, but those sources are low-quality, are mostly monochrome, and would generally make for a poor swap. If we could improve that data enough to train a model, we could then bootstrap to create a whole new view of Albert Einstein.

In particular, domain-specific tools will be beneficial to deepfakes. There are already face-specific upscalers that can be used on Faceswap's output to increase the effective resolution of the deepfakes process. Especially when used with very high-quality videos, upscaling the output can bridge the gap between computational power and the resolution of the swapped face.

Text generation

This one seems like it might be the least important for deepfakes, and in a way, the "generation" part is, yes. But techniques and technologies such as CLIP could be brought into deepfakes in new and unique ways. For example, what if we got rid of the general-purpose face encoder that is a part of all deepfakes and replaced it with a new encoder that was trained like CLIP, contrasting with various faces and data? We might be able to cut one of the largest parts of the model to run separately, allowing us to focus all our energy (and compute) on the decoder that builds the face back instead of wasting time and energy training an encoder just to give us an embedding to feed into the decoder.

The future of deepfakes

I think that with all the innovations that have come to generative AI recently, we can be sure that deepfakes have not reached the end of the line. There are innovations in these other domains that are just waiting to be implemented in deepfakes. These improvements will allow deepfakes to do new and interesting things while still keeping the essence of what a deepfake is – the swapping of one face with another.

We've not yet seen the end of deepfakes in movies. In fact, it seems likely that deepfakes in movies will only grow as the technology becomes more mature, capable, and available. It seems reasonable that entire actors may be deepfaked in the future just to keep a character from changing due to growing older or passing away. Imagine a sequel to *Gone with the Wind* done entirely with the original actors animated by AI under the complete direction of the director. It's feasible that we may see that before too long if AI continues to grow and advance.

Deepfakes are not unbounded; there are still problems to solve, such as resolution, training time, and even data collection. But it's now possible to make a deepfake face in a resolution higher than early Blu-rays with data collected from black-and-white movies, colorized and cleanly upscaled. Where will we be in 10 years? Or even 20 years? How long until you don't even bother to get out of bed to attend a video call where your full body is generated and matched to you?

The future of AI ethics

Our guidelines from *Chapter 3, Examining Deepfake Ethics and Dangers*, are still a solid foundation for ethics (after all, being nice to people never goes out of style), but as we move into the future, we'll approach entirely new challenges that will need their own ethical guidelines.

When, for example, is it acceptable to replace an actor with a deepfake? Is it acceptable to replace a hard-working human with an AI that is just puppeted by the director? What if it's a role that nobody wants to (or can) play? *Gollum* in Peter Jackson's *Lord of the Rings* was played by the incredibly talented character actor *Andy Serkis* (Serkis also played Snoke in Disney's *Star Wars* sequels and many other digital-first characters). Would it be ethical to replace him with an AI?

Chances are that an entirely AI character would be considered ethical by many (probably not the actors, though). But what if a director decided that an actress's voice wasn't "girly" enough and replaced it with a squeaky valley girl's voice without the actress's permission? If it's the director's creative vision to replace the actor playing the bad guy with a darker-skinned person? What about the example in the previous section where you replace your pajamas with a well-coordinated outfit for your video call? These questions are harder to answer, and it's something that we'll have to evaluate together as society moves forward.

Summary

Generative AI has a huge history and a tremendous future. We're standing before a vast plane where anything is possible, and we just have to go toward it. That said, not everything is visible today, and we must temper our expectations. The main challenges are the limitations of our computers, time, and research. If we dedicate our time and efforts to solving some of AI's limitations, we'll inevitably come up with brand-new leaps that will help us move forward. Even without huge revolutionary improvements though, there are a lot of smaller evolutionary improvements that we can make to improve the capabilities of these models.

The biggest driver of innovation is need. Having more and more people using generative AI and putting it toward novel uses will create the economic and social pushes that generative AI needs to continue being improved on into the future.

This book has all been about getting us to this point where we, the authors, can invite you to go forward and help AI move toward this generative future that, hopefully, we can all see just over that horizon.

Index

www.packtpub.com

Subscribe to our online digital library for full access to over 7,000 books and videos, as well as industry leading tools to help you plan your personal development and advance your career. For more information, please visit our website.

Why subscribe?

- Spend less time learning and more time coding with practical eBooks and Videos from over 4,000 industry professionals

- Improve your learning with Skill Plans built especially for you

- Get a free eBook or video every month

- Fully searchable for easy access to vital information

- Copy and paste, print, and bookmark content

Did you know that Packt offers eBook versions of every book published, with PDF and ePub files available? You can upgrade to the eBook version at packt.com and as a print book customer, you are entitled to a discount on the eBook copy. Get in touch with us at customercare@packtpub.com for more details.

At www.packt.com, you can also read a collection of free technical articles, sign up for a range of free newsletters, and receive exclusive discounts and offers on Packt books and eBooks.

Other Books You May Enjoy

If you enjoyed this book, you may be interested in these other books by Packt:

Network Science with Python

David Knickerbocker

ISBN: 978-1-80107-369-1

- Explore NLP, network science, and social network analysis
- Apply the tech stack used for NLP, network science, and analysis
- Extract insights from NLP and network data
- Authenticate and scrape tweets, connections, the web, and data streams
- Discover the use of network data in machine learning projects

Graph Data Science with Neo4j

Estelle Scifo

ISBN: 978-1-80461-274-3

- Use the Cypher query language to query graph databases such as Neo4j
- Build graph datasets from your own data and public knowledge graphs
- Make graph-specific predictions such as link prediction
- Explore the latest version of Neo4j to build a graph data science pipeline
- Run a scikit-learn prediction algorithm with graph data
- Train a predictive embedding algorithm in GDS and manage the model store

Packt is searching for authors like you

If you're interested in becoming an author for Packt, please visit `authors.packtpub.com` and apply today. We have worked with thousands of developers and tech professionals, just like you, to help them share their insight with the global tech community. You can make a general application, apply for a specific hot topic that we are recruiting an author for, or submit your own idea.

Share your thoughts

Now you've finished *Exploring Deepfakes*, we'd love to hear your thoughts! Scan the QR code below to go straight to the Amazon review page for this book and share your feedback or leave a review on the site that you purchased it from.

`https://packt.link/r/1-801-81069-9`

Your review is important to us and the tech community and will help us make sure we're delivering excellent quality content.

Download a free PDF copy of this book

Thanks for purchasing this book!

Do you like to read on the go but are unable to carry your print books everywhere?

Is your eBook purchase not compatible with the device of your choice?

Don't worry, now with every Packt book you get a DRM-free PDF version of that book at no cost.

Read anywhere, any place, on any device. Search, copy, and paste code from your favorite technical books directly into your application.

The perks don't stop there, you can get exclusive access to discounts, newsletters, and great free content in your inbox daily

Follow these simple steps to get the benefits:

1. Scan the QR code or visit the link below

https://packt.link/free-ebook/9781801810692

2. Submit your proof of purchase
3. That's it! We'll send your free PDF and other benefits to your email directly

www.ingramcontent.com/pod-product-compliance
Lightning Source LLC
Chambersburg PA
CBHW060130060326
40690CB00018B/3824